# Glossary of Research Methodology

NIPA GENX ELECTRONIC RESOURCES & SOLUTIONS P. LTD.
New Delhi-110 034

# Glossary of Research Methodology

**Awadhesh Kishore** Ph.D.
Professor
Department of Animal Sciences
School of Agriculture
ITM University
Gwalior, Madhya Pradesh

**NIPA GENX ELECTRONIC RESOURCES & SOLUTIONS P. LTD.**
New Delhi-110 034

**NIPA GENX ELECTRONIC**
**RESOURCES & SOLUTIONS P. LTD.**

101,103, Vikas Surya Plaza, CU Block
L.S.C.Market, Pitam Pura, New Delhi-110 034
Ph : +91 11 27341616, 27341717, 27341718
E-mail:newindiapublishingagency@gmail.com
www: www.nipabooks.com

***For customer assistance, please contact***
Phone: + 91-11-27 34 17 17
Fax: + 91-11- 27 34 16 16
E-Mail: feedbacks@nipabooks.com



ISBN: 978-93-95319-74-4



Composed and Designed by NIPA.

# Preface

In this book an attempt to clearly illustrate the terms related to the aspect of social research in the context of Extension Education. In order to ease the understanding of basic concepts, widely used simplest research methods and tools and techniques of data collection, we have divided the book into four parts, Foundations of social research, Research methods, Tools and techniques of data collection and Data processing and report writing.

Research Methods deals with the three major research methods used in extension education/ Agricultural extension, namely survey research, action research and case study.

We hope that, this book will be instrumental in developing the understanding of scientific research and its process among the students of agricultural extension/ extension education.

**Author**

# Contents

# A

**Abbreviations:** Shortened form of a word or a term is termed abbreviation.

**Abstract:** An abstract is a brief summary of a research article, thesis, review, conference proceeding, or any in-depth analysis of a particular subject and is often used to help the reader quickly ascertain the paper's purpose. The abstract must be better-worded, most concise, and most appealing part of the research proposal including a brief statement of the problem, objectives, procedures, materials, methods, achievements, benefits, funding requirements, research workers' and organizations' capability and credibility to carry-out research.

**Accidental sampling:** Accidental sampling as quota sampling is based upon the convenience in accessing sampling population. While quota sampling attempts comprise people possessing an obvious/visible characteristic, accidental sampling makes no such effort. Any person that comes across can be contacted for participation in the study. Collecting data are stopped when reaching the required number of respondents decided to have in the sample.

**Acronym:** An acronym is a word appeared from the initial letters of other words, for example, FAO (Food and Agriculture Organization).

**Action research:** The process by which practitioners attempt to study their problems scientifically in order to guide, correct and evaluate their decision and action is what a number of people have called action research. Action research imultaneously assists in practical problem-solving and expands scientific knowledge, as well as enhancing the competencies of the respective actors, being performed collaboratively in an immediate situation using data feedback in a cyclical process aiming at an increased understanding of a given situation, primarily applicable for the understanding of change processes in social systems and undertaken within a mutually acceptable ethical framework. In common with participatory research and collaborative inquiry, is based upon a philosophy of community development that seeks the involvement of community members in planning, undertaking, developing and implementing research and programme agendas. Research is a means to action to deal with a problem or an issue confronting a group or community. It follows a cyclical process that is used to identify the issues, develop strategies and implement the programmes to deal with them and then again assessing strategies in light of the issues. Action research is organized, investigative activity, aimed towards to study and constructive change of given endeavour by individual or group concerned with change and improvement.

**Active variable:** In studies that seek to establish causality or association there are variables that can be changed, controlled and manipulated either by a research worker or by someone else. Such variables are called active variables.

**Additive property of $\chi^2$:** Additive property of $\chi^2$ means that several values of $\chi^2$ can be added together. If the degrees of freedom are also added, this number gives the degrees of freedom for a total value of $\chi^2$. Thus, a number of $\chi^2$ values have been obtained from a number of samples of similar data, then the various values of $\chi^2$ can be combined by just simply adding them.

**After-only design:** In an after-only design the research worker knows that a population is being, or has been, exposed to an intervention and wishes to study its impact on the population. In this design, baseline information (pre-test or before observation) is usually 'constructed' either on the basis of respondents' recall of the situation before the intervention, or from information available in existing records, i.e. secondary sources.

**Alternative hypothesis:** The formulation of an alternative hypothesis is a convention in scientific circles. Its main role is to specify explicitly the relationship that will be regarded as true in case the research hypothesis proves to be wrong. In a way, an alternative hypothesis is the opposite of the research hypothesis. In other words, if results do not support null hypothesis, and it is concluded that something else is true, that is known as alternative hypothesis, and symbolically it is expressed as $H_{\alpha}$: $\mu \neq \mu_{H0}$.

**Ambiguous question:** An ambiguous question is one that contains more than one meaning and that can be interpreted differently by different respondents.

**Analogue computer:** The analogue computer is one which is operated by measuring rather than counting.

**Analysis of co-variance:** While implementing the ANOCOVA technique, the impact of uncontrolled variable is frequently removed by a simple linear regression technique and the residual sums of squares are used to provide variance estimates which in turn are used to make tests of significance.

**Analysis of variance:** Analysis of variance (abbreviated as ANOVA) is an enormously useful technique concerning researches several disciplines. This technique is used when multiple sample cases are involved.

**Analytical research:** The analytical research is based on the analysis of facts or information already available, to make a critical evaluation of the material and to draw valid conclusions.

**Annual reports:** Annual reports include straight forward descriptions of work that has been done during a year or 12-month period. The intention is not so much to prove a hypothesis, rather to describe activities, justify budget expenditure in terms of research undertaken, and demonstrates the impact to attract more funding.

**ANOCOVA:** *See Analysis of co-variance.*

**ANOVA:** *See Analysis of variance.*

**Application software:** Application software is that program which tells the computer how to perform specific tasks such as preparation of company pay-roll or inventory management.

**Applied research:** Applied research is a kind of research aimed to find out a solution of immediate problem facing a society or an industrial/business organization. Applied research is one where research techniques, procedures and methods that form the body of research methodology are applied to collect information about various aspects of a situation, issue, problem or phenomenon so that the information gathered can be utilized for other purposes such as policy formulation, programme development, programme modification and evaluation, enhancement of the understanding about a phenomenon, establishing causality and outcomes, identifying needs and developing strategies.

**Arbitrary approach of scaling:** Arbitrary approach of scaling is an approach where scale is developed on an ad hoc basis.

**Area chart:** For variables measured on an interval or a ratio scale. Information about the sub-categories of a variable can also be presented in the form of an area chart. It is plotted in the same way as a line diagram with the area under each line shaded to highlight the magnitude of the subcategory in relation to other subcategories. Thus an area chart displays the area under the curve in relation to the subcategories of a variable.

**Area sampling:** Area sampling is quite close to cluster sampling and is often talked about when the total geographical area of interest happens to be a big one. Under area sampling, the total area is divided into a number of smaller non-overlapping areas, generally called geographical clusters, then a number of these smaller areas are randomly selected, and all units in the small areas are involved in the sample. Area sampling is especially helpful where the list of the population concerned is not available.

**Attitudinal scales:** Those scales that are designed to measure attitudes towards an issue are called attitudinal scales. There are three types of scale:

the summated rating scale (Likert scale), the equal-appearing interval scale (Thurstone scale) and the cumulative scale (Guttmann scale).

**Attitudinal score:** A number that has been calculated having assigned a numerical value to the response given by a respondent to an attitudinal statement or question. Different attitude scales have different ways of calculating the attitudinal score.

**Attitudinal value:** An attitudinal scale comprises many statements reflecting attitudes towards an issue. The extent to which each statement reflects this attitude varies from statement to statement. Some statements are more important in determining the attitude than others. The attitudinal value of a statement refers to the weight calculated or given to a statement to reflect its significance in reflecting the attitude: the greater the significance or extended, the greater the attitudinal value or weight.

**Attribute variables:** The variables that cannot be manipulated, changed or controlled, and that reflect the characteristics of the study population, for example, age, gender, education and income etc.

**Attributes:** The attributes are object of study are referred to as characteristics and the units possessing them. The attributes or qualitative phenomena are also quantified on the basis of the presence or absence of the concerning attribute(s). In science and research, attribute is a characteristic of an object (person, thing, etc.).

# B

**Bar chart:** The bar chart or diagram is one of the ways of graphically displaying categorical data. A bar chart is identical to a histogram, except that in a bar chart the rectangles representing the various frequencies are spaced, thus indicating that the data is categorical. The bar diagram is used for variables measured on nominal or ordinal scales.

**The bar diagram:** *See Bar chart.*

**Before-and-after studies:** A before-and-after design can be described as two sets of cross-sectional data collection points on the same population to find out the change in a phenomenon or variable(s) between two points in time. The change is measured by comparing the difference in the phenomenon or variable(s) between before and after observations.

**Bias:** Bias is a deliberate attempt either to conceal or highlight something that is supposed to be found in the research or to use deliberately a procedure, or method that is not appropriate but will provide information that is looked for because of vested interest in it.

**Bibliography:** Bibliography is the complete list of all the references appeared within the text, whether they appear within the text or not. The bibliography, which is generally appended to the research report, is an alphabetically or chronologically arranged list of books, research papers and other sources of literature which are referred in the report.

**Binary digits:** Computers use only the binary digits, which is a system in which all the numbers are represented by a combination of two digits viz. one and zero.

**Binary number system:** An arithmetic concept which comprises two levels, instead of ten, but operates on the same logic is called the binary system.

**Bivariate population:** Bivariate population is the population consisting of measurement of only two variables. In other words if for every measurement of a variable, X, has a corresponding value of a second variable, Y, the resulting pairs of values are called a bivariate population.

**Blind studies:** In a blind study, the population does not know whether it is getting real or fake treatment or which treatment modality in the case of comparative studies. The main objective of designing a blind study is to isolate the placebo effect.

**Blogs:** A blog or discussion group (a contraction of the term 'web log') is a type of website, usually maintained by an individual, with regular entries of commentary, descriptions of events, or other material such as graphics or video. Blogs are personal, representing the views of the writer,

**Book-reviews:** Book-reviews are the analysis the content of the book and reporting on the author's intentions, his success or failure in finding his aims, his language, his style, scholarship, bias or his point of view.

**Books and book chapters:** The book and book chapter is a synthesis of knowledge and information about a particular subject. It is more likely to form one part of an overall contents list that, taken together, exhaustively describes a clearly defined aspect of one field.

# C

**Canonical analysis:** The canonical analysis can be used in the case when both measurable and non-measurable variables for the purpose of simultaneously predicting a set of dependent variables from their joint covariance with a set of independent variables.

**Canonical correlation analysis:** The canonical correlation analysis technique was first developed by Statistician Hotelling. It is an effort to simultaneously predict a set of criterion variables from their joint co-variance with a set of explanatory variables. Both the types of data, i.e. metric and non-metric, can be utilized in this multivariate technique.

**Car-relational studies:** Studies, which are primarily designed to investigate whether or not there is a relationship between two or more variables, are called cor-relational studies.

**Case of tying ranks:** The average method of assigning ranks is adopted if tied ranks are occurring. If the ties are not numerous, 'W' may be calculated as stated in Spearman's rank correlation coefficient method, without making any adjustment in the formula. But if the ties are numerous, The correction factor is calculated for each set of ranks.

**Case study:** The case study method is a technique by which individual factor whether it is an institution or just an episode in the life of an individual or a group is analyzed in its relationship to any other in the group. The case study design is based upon the assumption that the case being studied is atypical of cases of a certain type and therefore a single case can provide insight into the events and situations prevalent in a group from where the case has been drawn. In a case study design the 'case' selected, becomes the basis of a thorough, holistic and in-depth exploration of the aspect(s) that is required to find out about. It is an approach in which a particular instance or a few carefully selected cases are studied intensively. To be called a case study it is important to treat the total study population as one entity. It is one of the important study designs in qualitative research.

**Categorical scales:** Categorical scales are also known as rating scales.

**Categorical variables:** Categorical variables are those variables where the unit of measurement is in the form of categories. On the basis of the presence or absence of a characteristic, a variable is placed in a category. There is no measurement of the characteristics as such. In terms of measurement scales such variables are measured in nominal or ordinal scales. Rich/poor, high/low, hot/cold are the examples of categorical variables.

**Causal analysis:** Causal analysis is concerned with the study of how one or more variables affect changes in another variable. It is considered relatively more important in experiments where as in most of surveys the interest lies in understanding and controlling relationships between variables and correlation analysis is relatively more important in surveys.

**Central editing of data:** The central editing of data happens when all structures or schedules questionnaires, perception or interview have been finished and come back to the workplace.

**Central limit theorem:** Central limit theorem explains the sort of relationship between the shape of the population distribution and the sampling distribution of the mean.

**Centroid Method of Factor Analysis:** The Centroid method of factor analysis, was developed by Thurstone. The centroid method tends to maximize the sum of loadings, disregarding signs.

**Chance sampling:** The chance sampling or Simple random sampling or probability sampling means a method in which each and every item present in the population has an equal chance of inclusion in the sample and each one of the possible samples, in case of a finite universe, has the similar probability of being selected. This is the most commonly used methods of selecting a random sample. It is a process of selecting the required sample size of the sampling population, providing each element with an equal and independent chance of selection by any method designed to select a random sample.

**Chance variable:** In studying causality or association there are times when the mood of a respondent or the wording of a question can affect the reply given by the respondent when asked again in the post-test. There is no systematic pattern in terms of this change. Such variables are called chance or random variables. When collecting information from respondents, there are times when the mood of a respondent or the wording of a question can affect the way a respondent replies. There is no systematic pattern in terms of this change. Such shifts in responses are said to be caused by random or chance variables.

**Charles Spearman's coefficient:** Charles Spearman's coefficient of correlation (also a rank correlation) is the technique to determine the degree of correlation between two variables in case of ordinal data where ranks are given to the different values of the variables.

**Chi-square ($\chi^2$) distribution:** Chi-square distribution is encountered in the condition when dealt with collections of values that involve adding up into squares. Variances of samples require adding a collection of squared quantities and thus having distributions that are related to chi-square distribution.

**Chronological development:** Chronological development is anchored in a connection or sequence in time or occurrence.

**Classes in the final order:** Classes in the final order are those classes developed on the basis of attributes or variables considered.

**Classification of data:** Classification of data is a process in which data having a common characteristic are placed in one class and in this way the entire data get divided into a number of groups or classes.

**Client-oriented evaluation:** The core philosophy of consumer-oriented evaluation or client-oriented evaluation rests on the assumption that assessment of the value or merit of an intervention–including its effectiveness, outcomes, impact and relevance–should be judged from the perspective of the consumer. Consumers, according to this philosophy, are the best people to make a judgment on these aspects. An evaluation done within the framework of this philosophy is known as consumer-oriented evaluation or client-centered evaluation.

**Clinical interview:** The clinical interview is concerned with broad underlying feelings or motivations or with the course of individual's life experience. The method of eliciting information under it is generally left to the interviewer's discretion.

**Closed question:** In a closed question the possible answers are set out in the questionnaire or interview schedule and the respondent or the investigator ticks the category that best describe a respondent's answer.

**Cluster Analysis:** Cluster analysis includes methods of classifying variables into clusters.

**Cluster sampling and area sampling:** Cluster sampling involves grouping the population called clusters and then selecting the clusters rather than individual elements for inclusion in the sample. Cluster sampling is based on the ability of the research worker to divide a sampling population into groups (based upon a visible or easily identifiable characteristic), and then select elements from each cluster using the SRS technique. Clusters can be formed on the basis of geographical proximity or a common characteristic that has a correlation with the main variable of the study (as in stratified sampling). Depending on the level of clustering, sometimes sampling may be done at different levels. These levels constitute the different stages (single, double or multiple) of clustering.

**Code:** The numerical value that is assigned to a response at the time of analyzing the data.

**Codebook:** A listing of a set of numerical values (set of rules) that you decided to assign to answers obtained from respondents in response to each question is called a code book.

**Coding operation:** Coding operation is an action that is usually done at this stage through which the categories of data are transformed into symbols that may be tabulated and counted.

**Coding:** Coding means to the process of assigning numerals or other symbols to answer so that responses can put into a limited number of categories or classes. In other words, the process of assigning numerical values to different categories of responses to a question for the purpose of analyzing them is called coding. Such classes should be appropriate to the research problem under consideration.

**Coefficient of the standard deviation:** If the standard deviation is divided by the arithmetic average of the data series, the resulting quantity is known as the coefficient of standard deviation.

**Coefficient of variation:** If the coefficient of standard deviation is multiplied by 100, the resulting figure is known as the coefficient of variation.

**Cohort studies:** The cohort studies are based upon the existence of a common characteristic such as year of birth, graduation or marriage, within a subgroup of a population that is needed to study. People with the common characteristics are studied over a period of time to collect the information of interest of the research. Studies could cover the fertility behaviour of women born in 1986 or career paths of 1990 graduates from a medical school, for instance. Cohort studies look at the trends over a long period of time and collect data from the same group of people.

**Collaborative inquiry:** The collaborative inquiry is another name for participatory research that advocates a close collaboration between the research worker and the research participants.

**Column percentages:** The column percentages are calculated from the total of all the subcategories of one variable that are displayed along a column in different rows.

**Commonality ($h_2$):** Communality, represented as $h_2$, reports how much of each variable is represented by the basic factor taken together. A high estimation of commonality trait implies that there is little of the variables are left finished after whatever the factors represent is taken into consideration.

**Communication:** Communication can be defined as to give to another, impart, transmit, make known; to have an interchange of thoughts.

**Community discussion forum:** A community discussion forum is a qualitative strategy designed to find opinions, attitudes, ideas of a community with regard to community issues and problems. It is one of the very common ways of seeking a community's participation in deciding about issues of concern to it.

**Comparative scale:** Comparative scale is used when a respondent scores some object without direct reference to other objects. Under comparative scales, which are also known as ranking scales, the respondent is asked to compare two or more objects. In this sense the respondent may state that one object is superior to the other or those three models of pen rank in order 1, 2 and 3.

**Comparative study design:** Sometimes it is needful to compare the effectiveness of different treatment modalities. In such situations a comparative design is used. With a comparative design, as with most other designs, a study can be carried out either as an experiment or non-experiment. In the comparative experimental design, the study population is divided into the same number of groups as the number of treatments to be tested. For each group the baseline with respect to the dependent variable is established. The different treatment modalities are then introduced to the different groups. After a certain period, when it is assumed that the treatment models have had their effect, the 'after' observation is carried out to ascertain changes in the dependent variable.

**Completely randomized design (C.R.D.):** Completely randomized design involves only two principles viz., the principle of replication and the principle of randomization of experimental designs.

**Complex factorial design:** When a factorial experiment is done with more than two factors, complex factorial design is used.

**Complex sample design:** The complex sample design is somewhat a sequential sampling where the ultimate size of the sample is not fixed in advance, but is determined according to mathematical decisions on the basis of information yielded as the survey progresses.

**Complex tabulation:** The complex table shows the division of data in two or more categories and thus, is designed to give information concerning one or more sets of interrelated questions.

**Composite hypothesis:** In case a hypothesis is not simple or specific, but if it is of the type $\mu \neq \mu_{H0}$ or $\mu > \mu_{H0}$ or $\mu < \mu_{H0}$, then the hypothesis are called as a composite (or nonspecific) hypothesis.

**Computer:** Computer is a machine capable to receive, store, edit, manipulate and yield information such as numbers, words, pictures, audios and videos.

**Computer systems:** Computer systems can be described as containing some kind of input devices, the CPU and some kind of output devices.

**Concept:** In defining a research problem or the study population it is needful to use certain words that as such are difficult to measure and/or the understanding of which may vary from person to person, these words are called concepts. In order to measure them they need to be converted into indicators (not always) and then variables. Words like satisfaction, impact, young, old, and happy etc. are concepts as their understanding would vary from person to person.

**Concept development:** Concept development means that the research worker should arrive at an understanding of the major concepts pertaining to his study. This step is more apparent in theoretical studies than in the more pragmatic research, where the fundamental concepts are often already established.

**Conceptual framework:** A conceptual framework stems from the theoretical framework and concentrates, usually, on one section of that theoretical framework which becomes the basis of the research study. The latter consists of the theories or issues in which the research study is embedded, whereas the former describes the aspects selected from the theoretical framework to become the basis of the research inquiry. The conceptual framework is the basis of the research problem.

**Conceptual research:** Conceptual research is the research that is related to some abstract idea(s) or theory. It is generally used to develop new concepts or to reinterpret existing ones.

**Concurrent validity:** The concurrent validity refers to the usefulness of a test in closely relating to other measures of known validity. When it is needful to investigate how good a research instrument is by comparing it with some observable criterion or credible findings, this is called concurrent validity. It is comparing the findings of the research instrument with those found by another which is well accepted. Concurrent validity is identified by how well an instrument compares with a second evaluation done concurrently.

**Conditioning effect:** The conditioning effect describes a situation where, if the same respondents are contacted frequently, they begin to know what is expected of them and may respond to questions without thought, or they may lose interest in the inquiry, with the same result. This situation's effect on the quality of the answers is known as the conditioning effect.

**Conference papers and posters:** Conferences offer scientists an opportunity to present results of research that is still at a preliminary stage, but that contains interesting developments. Because time is limited during a conference session, papers that are presented orally at conferences are necessarily short.

**Confidence level:** The confidence level or reliability level is the supposed assumed level in terms of the percentage of times that the real estimated value will lie within the expressed exactness limits.

**Confirmability:** The confirmability refers to the degree to which the results obtained through qualitative research could be confirmed or corroborated by others. Confirmability in qualitative research is similar for reliability in quantitative research.

**Confounded relationship:** When the dependent variable is not free from the influence of extraneous variable(s), the relationship between the dependent and independent variables is said to be confounded relationship.

**Consensus approach of scaling:** Consensus approach of scaling is a panel of judges evaluates the items chosen for inclusion in the instrument in terms of whether those are relevant to the topic area and unambiguous in implication.

**Constant variable:** When a variable can have only one category or value, for example taxi, tree and water, it is known as a constant variable.

**Construct validity:** A measure is construct validity to the extent that it verifies to anticipate correlations with other theoretical propositions. Construct validity is how much scores on a test can be represented by the explanatory constructs of a sound theory. This type of validity is a more sophisticated technique for establishing the validity of an instrument, which is based upon statistical procedures. It is determined by ascertaining the contribution of each construct to the total variance observed in a phenomenon.

**Consumer-oriented evaluation:** *See Client-oriented evaluation.*

**Consumer panels:** An additional room of the pantry audit approach on a regular basis is known as consumer panel, where a set of consumers is organized to come to an understanding to maintain featured daily records of their consumption and the same is made available to investigators on demands.

**Content analysis:** Content analysis is one of the main methods of analyzing qualitative data. It is the process of analyzing the contents of interviews or observational field notes in order to identify the main themes that emerge from the responses given by the respondents or the observation notes made by a research worker.

**Content validity:** Content validity is the extent to which a measuring instrument provides an adequate coverage of the topic under study. If the instrument encloses a representative sample of the universe, the content validity is fine. Its determination is principally judgmental and intuitive. It can also be decided by using a panel of persons who shall judge how well the measuring tool meets the standards, but there is no numerical way to express it. In addition to linking each question with the objectives of a study as a part of establishing the face validity, it is also important to examine whether the questions or items have covered all the areas of interest to cover in the study. Examining questions of a research instrument to establish the extent of coverage of the areas under study is called content validity of the instrument.

**Content-analysis:** Content-analysis includes analyzing the contents of documentary materials such as books, magazines, newspapers and the contents of all other verbal materials which can be either spoken or printed.

**Continuing consumer panel:** A continuing consumer panel is often set up of consumer panel for an indefinite period with a view to collect data on a particular aspect of consumer behaviour over time, generally at periodic intervals on a variety of subjects.

**Continuous variables:** Continuous variables have continuity in their unit of measurement. In other words, phenomena which can take on quantitative different values even in decimal points are called as continuous variables. A continuous variable is that which can assume any numerical value within a specific range; for example age, income and attitude score. Those can take on any value of the scale on which they are measured. Age can be measured in years, months and days. Similarly, income can be measured in dollars and cents.

**Control design:** In experimental studies that aim to measure the impact of an intervention, it is important to measure the change in the dependent variable that is attributed to the extraneous and chance variables. To quantify the effect of these sets of variables another comparable group is selected that is not subjected to the intervention. Study designs where a control group isolates the impact of extraneous and change variables are called control design studies.

**Control group:** The group in an experimental study which is not exposed to the experimental intervention is called a control group. The sole purpose of the control group is to measure the impact of extraneous and chance variables on the dependent variable. In other words, in an experimental hypothesis, -testing research when a group is exposed to the usual conditions, is termed a control group.

**Controlled observation:** The controlled observation is the observation when it takes place according to definite pre-arranged plan, involving experimental procedure.

**Convenience sampling:** When populace components are chosen for incorporation in the sample based on the ease of access, it can be called convenience sampling.

**Convenience test:** Convenience test suggests that the measuring instrument should be easy to administer. For this purpose due attention is required to the proper layout of the measuring instrument. For instance, a questionnaire, with clear instructions (illustrated by examples), is certainly more effective and easier to complete than one which lacks these features.

**Correlation analysis:** Correlation analysis studies the joint variation of two or more variables for determining the amount of correlation between two or more variables.

**Cost–benefit evaluation:** The central aim of a cost–benefit evaluation is to put a price tag on an intervention in relation to its benefits.

**Cost-effectiveness evaluation:** The central aim of a cost-effectiveness evaluation is to put a price tag on an intervention in relation to its effectiveness.

**Credibility:** Credibility in qualitative research is parallel to internal validity in quantitative research and refers to a situation where the results obtained through qualitative research are agreeable to the participants of the research. It is judged by the extent of respondent concordance whereby a research worker takes his findings to those who participated in his research for confirmation, congruence, validation and approval: the higher the outcome of these, the higher the credibility (validity) of the study.

**Criterion-related validity:** Criterion-related validity is expressed as the coefficient of correlation between test scores and some measure of future performance or in other words, between test scores and scores on another measure of known validity. This type of validity relates to ability to predict some outcome or estimate the existence of some current condition. This form of validity reflects the success of measures used for some empirical estimating purposes.

**Crossover comparative experimental design:** In the crossover design, also called the ABAB design, two groups are formed, the intervention is introduced to one of them and, after a certain period, the impact of this intervention is measured. Then the interventions are 'crossed over'; that is, the experimental group becomes the control and vice versa.

**Cross-sectional studies:** Cross-sectional studies, also known as one-shot or status studies are the most commonly used designs in the research work. This design is the best suited to studies aimed at finding out the prevalence of a phenomenon, situation, problem, attitude or issue, by taking a cross-section of the population. Those are useful in obtaining an overall 'picture' as it stands at the time of the study.

**Cross-tabulation:** Cross-tabulation is a statistical procedure that analyses two variables, usually independent and dependent or attribute and dependent, to determine if there is a relationship between them. The subcategories of both the variables are cross-tabulated to ascertain if a relationship exists between them.

**Cumulative frequency polygon:** The cumulative frequency polygon or cumulative frequency curve is drawn on the basis of cumulative frequencies. The main difference between a frequency polygon and a cumulative frequency polygon is that the former is drawn by joining the midpoints of the intervals, whereas the latter is drawn by joining the end points of the intervals because cumulative frequencies interpret data in relation to the upper limit of an interval.

**Cumulative scale:** Cumulative scale, also called Louis Guttmann's scalogram analysis, comprises of a series of statements to which a respondent communicates his agreement or contradiction. The special element of this kind of scale is that statement in the scale form a cumulative series.

# D

**Data Analysis:** In the process of data analysis, connections or differences supporting or clashing with unique or new hypotheses should be subjected to statistical tests of significance to decide to what validity data can be said to show any conclusions.

**Defining a problem:** The term defining a problem means "To pin-point the problem or defining a problem to reach the core of the problem i.e. threadbare analysis."

**Deliberate sampling:** Deliberate sampling or judgmental sampling or purposive or non-probability sampling technique involves object based or deliberate selection of specific units of the universe for constituting a sample which represents the universe. The primary consideration in this sampling design is the judgment as to who can provide the best information to achieve the objectives of the research study. The research worker only go to those people who in his opinion are likely to have the required information and are willing to share it with the research worker.

**Dependability:** Dependability in qualitative research is very similar to the concept of reliability in quantitative research. It is concerned with whether obtained results are the same results if could observe the same thing twice: the greater the similarity in two results, the greater the dependability.

**Dependent variable:** When establishing causality through a study, the variable assumed to be the cause is called an independent variable and the variables in which it produces changes are called the dependent variables. A dependent variable is dependent upon the independent variable and it is assumed to be because of the changes.

**Depth interviews:** Depth interviews are detailed interviews that are designed to discover underlying motives and desires. Such interviews are held to explore needs, desires and feelings of the respondents.

**Descriptive characteristics of data:** Descriptive characteristics of the data refer to a qualitative phenomenon which cannot be measured quantitatively; only their presence or absence in an individual item can be noticed.

**Descriptive research studies:** Descriptive research study involves surveys and fact-finding different types of injuries. This type of research studies are the studies that are concerned with describing the traits of a particular individual, or of a group. In other words, a study in which the main focus is given on description, rather than examining relationships or associations, is classified as a descriptive research study. A descriptive research study attempts systematically to describe a situation, problem, phenomenon, service or programme, or provides information about, for example, the living conditions of a community, or describes attitudes towards an issue.

**Develop indicators:** Develop indicators for measuring each concept element. Indicators are specific questions, scales, or other devices by which respondent's knowledge, opinion, expectation, etc., is measured.

**Diagnostic research:** Diagnostic research studies explain the frequency with which something occurs or its association with something else.

**Dichotomous variable:** When a variable can have only two categories as in male/female, yes/no, good/bad, head/tail, up/down and rich/poor, it is known as a dichotomous variable.

**Differential Scale**: The differential scale is also called Thurston-type Scales. In this scale an approach of the selection of items is made by a panel of judges who evaluate the items in terms of whether they are relevant to the topic area and unambiguous in implication.

**Digital computer:** A digital computer is one which is operated essentially by counting (using information, including letters and symbols, in coded form).

**Discrete variables:** If the variables that can only be expressed in integer values, they are non-continuous variables or discrete variables. A variable for which the individual values fall on the scale only with distinct gaps is called a discrete variable.

**Discussion groups:** *See Blogs.*

**Disguised observation:** The disguised observation is the observation when the observer is observing such a manner that his presence may be unknown to the people whom he is observing.

**Dispersion:** Statistical devices which measure scatter of values of items of a variable in the data series around the true value of average are called dispersion.

**Disproportionate stratified sampling:** When selecting a stratified sample if it includes selecting an equal number of elements from each stratum without giving any consideration to its size in the study population, the process is called disproportionate stratified sampling.

**Dissertation:** The dissertation or thesis is written evidence of sustained research, testing a particular hypothesis in a novel area, made over a considerable period. It generally contains an extensive review of the literature, as well as the results of a number of experiments, testing a unified hypothesis. Some of the material or outcome may previously have been published in a series of research papers during the course of the research.

**Distribution-free test:** Distribution-free test is a test to test the hypothesis without a model.

**Distributor audits:** Distributor or store audits are carried out by distributors as well as manufacturing at regular intervals. Distributors get the retail stores audited and use such information for estimating market size, market share, seasonal purchasing pattern and so on.

**Double-barreled question:** A double-barreled question is a question within a question.

**Double-blind studies:** The concept of a double-blind study is very similar to that of a blind study except that it also tries to eliminate research worker bias by not disclosing to the research worker the identities of experimental, comparative and placebo groups. In a double-blind study neither the research worker nor the study participants know which study participants are receiving real, placebo or other forms of interventions. This prevents the possibility of introducing bias by the research worker.

**Double-control studies:** Although the control group design helps to quantify the impact that can be attributed to extraneous variables, it does not separate out other effects that may be due to the research instrument (such as the reactive effect) or respondents (such as the maturation or regression effects, or placebo effect). When it is needed to identify and separate out these effects, a double-control design is required. In a double-control study, includes two control groups instead of one. To quantify, say, the reactive effect of an instrument, it is needful to exclude one of the control groups from the 'before' observation.

# E

**Economy consideration:** Economy consideration suggests that some trade-off which is needed between the ideal research project and that which the budget can afford. The length of measuring tool is an important area where economic pressures are quickly felt. Even though more items give superior reliability, but in the interest of limiting the interview or observation time, the research worker is to take only a few items for his study purpose. Similarly, data-collection techniques to be utilized are also dependent at the times upon economic factors.

**Editing of data:** Editing of data is a process of examining the collected raw data, especially in surveys, to detect errors and omissions and to correct these when possible.

**Editing:** The editing is the procedure that improves the quality of the data for coding. In other words, editing is an activity of scrutinizing of completed research instruments, writings and/or documents; to identify and minimize errors, incompleteness, misclassification and gaps in the information obtained from respondents is also the part of editing.

**Eigen value:** When the sum of the squared values of factor loadings relating to a factor, is under consideration, then such sum is referred to as Eigen value or latent root. The Eigen value represents the relative importance of all the factors in accounting for the particular set of variables being analyzed.

**Elementary units:** *See attributes.*

**Elevation effect:** Some observers when using a scale to record an observation may prefer to use certain section(s) of the scale in the same way that some teachers are strict markers and others are not. When observers have a tendency to use a particular part(s) of a scale in recording an interaction, this phenomenon is known as the elevation effect.

**Empirical research:** Empirical research is the research that relies on experience or observation alone, often without due regard for system and theory. This type of research is data-based, coming up with conclusions that are capable of being verified by observations or experiments.

**Encyclopaedia:** Encyclopaedia can be defined as the source of concise information on a number of subjects written by specialists. It contains a convenient source of information, and often include illustrations and bibliographies. Only specialized encyclopaedia deals with restricted areas of knowledge.

**Endnote:** Endnote, given on the last page of the report, is meant for cross referencing, citing of authorities and sources, acknowledging and classifying, clarifying or explaining of a point of view but it is neither an end nor a means of the display of scholarship. Endnotes must be symbolized or numbered.

**Equivalence aspect:** The equivalence aspect considers how much error may get introduced by different investigators or different samples of the items being studied.

**Error of central tendency:** When using scales in assessments or observations, unless an research worker is extremely confident of his ability to assess an interaction, he may tend to avoid the extreme positions on the scale, using mostly the central part. The error this tendency creates is called the error of central tendency.

**Estimates:** The random variables, e.g. $\bar{X}$ and $\sigma^2_s$, used for estimation of population parameters, such as $\mu$ and $s_p^2$ are usually called as estimators.

**Estimators:** The specific values of the estimates of the population parameters, e.g. $\bar{X}$ and $\sigma^2_s$, are called estimators.

**Ethical practice:** Professional practice undertaken in accordance with the principles of accepted codes of conduct for a given profession or group is known as Ethical practice.

**Evaluation for planning:** Evaluation for planning addresses the issue of establishing the need for a programme or intervention.

**Evaluation:** Evaluation is a process that is guided by research principles for reviewing an intervention or programmes in order to make informed decisions about its desirability and/or identifying changes to enhance its efficiency and effectiveness.

**Evidence-based practice:** A service delivery system that is based upon research evidence as to its effectiveness; a service provider's clinical judgment as to its suitability and appropriateness for a client; and a client's preference as to its acceptance.

**Ex post facto research:** *See Descriptive research.*

**Experience survey:** Experience survey means the survey of people who have had practical experience with the problem to be studied. The object of such a survey is to obtain insight into the relationships between variables and new ideas relating to the research problem.

**Experiment:** The process of examining the facts of a statistical hypothesis, which is related to some research problem, is known as an experiment.

**Experimental approach to research:** It is characterized by much greater control over the research environment and in this case some variables are manipulated to observe their effect on other variables.

**Experimental Designs:** Experimental design refers to the framework or structure of an experiment and as such there are several experimental designs.

**Experimental error:** Whatever effect is noticed on the dependent variable as a result of extraneous is called as experimental error.

**Experimental group:** An experimental group is one that is exposed to the intervention, being tested to study its effects.

**Experimental groups:** In an experimental hypothesis-testing research when a group is exposed to some novel or special condition, it is termed an experimental group.

**Experimental research studies:** Experimental research studies or hypothesis-testing research studies (generally known as experimental studies) are those where the research worker tests the hypotheses of causal relationships between variables.

**Experimental research:** Experimental research is a systematic and logical method for answering the question, "What will happen if this is done when certain variables are carefully controlled or manipulated?"

**Experimental studies:** In studying causality, when a research worker or someone else introduces the intervention that is assumed to be the 'cause' of change and waits until it has produced–or has been given sufficient time to produce–the change, then in studies like this a research worker starts with the cause and waits to observe its effects. Such types of studies are called experimental studies.

**Experimental unit:** The pre-determined plots or the blocks, where different treatments are used, are known as experimental units. Such experimental units must be selected (defined) very carefully.

**Expert sampling:** Expert sampling is the selection of subjects with demonstrated or known expertise in the area of interest to become the basis of data collection. This type of sample is a group of experts from whom the required information is collected. It is like purposive sampling, where the sample comprises experts only.

**Explanatory research studies:** In an explanatory research study the main emphasis is to clarify why and how there is a relationship between two aspects of a situation or phenomenon. This type of research studies are termed as formulated research studies. The main objective of such studies is that to formulate a problem for more precise investigation or of expanding the working hypotheses from an operational attitude.

**Exploratory research studies:** The exploratory research studies are the research aimed to develop the hypotheses rather than their testing. This is when a study is undertaken with the objective either to explore an area where little is known or to investigate the possibilities of undertaking a particular research study. When a study is carried out to determine its feasibility it is also called a feasibility or pilot study.

**External validity:** External validity of research findings is their ability to generalize populations, settings, treatment variables and measurement variables.

**Extraneous variable:** Independent variables that are not related to the purpose of the study, but may affect the dependent variable are termed as extraneous variables. In studying causality, the dependent variable is the consequence of the change brought about by the independent variable. In everyday life there are many other variables that can affect the relationship between independent and dependent variables. These variables are called extraneous variables.

**Eye cameras:** Eye cameras are the device designed to record the focus of the eyes of a respondent on a specific portion of a sketch or diagram or written material.

# F

**F distribution:** Suppose, the variances of two independent samples, i.e. $(\sigma_{s1})^2$ and $(\sigma_{s2})^2$ are of size, i.e. $n_1$ and $n_2$, respectively, are taken from two independent normal populations, having the same variance, i.e. $(\sigma_{p1})^2 = (\sigma_{p2})^2$, the ratio F will be $(\sigma_{s1})^2/(\sigma_{s2})^2$. Where, $(\sigma_{s1})^2 = \Sigma(\bar{X}_{1i} - \bar{X}_1)^2 / n_1\text{-}1$ and $(\sigma_{s2})^2 = \Sigma(\bar{X}_{2i} - \bar{X}_2)^2 / n_2\text{-}1$ has an F distribution with degrees of freedom for two independent normal populations i.e. $n_1$-1 and $n_2$-1, respectively.

**Face validity:** When the inclusion of a question or item is justified in a research instrument by linking it with the objectives of the study, thus providing a justification for its inclusion in the instrument, the process is called face validity.

**Factor analysis:** Factor analysis is the most commonly employed multivariate procedure in the research studies when there is a systematic interdependence among a set of observed or manifest variables and the research worker is interested in finding out something more fundamental or latent which creates this commonality.

**Factor Scale:** Factor scale is explored through factor analysis or based on inter-correlations of items that indicate that a common factor account of the relationships between the items. This scale is particularly useful in uncovering latent attitude dimensions and approach scaling through the concept of multiple-dimension attribute space.

**Factor scores:** Factor score is defined as the degree to which each respondent obtains high scores on the group of items that load high on each factor.

**Factor:** A factor is an underlying dimension that explains several observed variables.

**Factorial designs:** Factorial designs are used in experiments where the effects of varying more than one factor are to be determined.

**Factor-loadings:** Factor-loadings are those values which make the explanations how closely the variables are related to each one of the factors discovered.

**Feasibility Report:** The feasibility report is submitted to the authorities that explore the feasibility of undertaking a particular project or not. It examines and compares alternatives, analyses the pros and cons, and suggests which, if any, of the alternatives are feasible.

**Feasibility study:** When the purpose of a study is to investigate the possibility of undertaking it on a larger scale and to streamlining methods and procedures for the main study, the study is called a feasibility study.

**Features:** Features, which are longer and more literary than news stories, are the most common format for science in popular publications. Most often, such articles are between 1000 and 3000 words long and the editor will expect you to keep to the number they request.

**Feminist research:** The feminist research is more concerned and theory acted as the guiding framework for this research. A focus on the viewpoints of women, the aim to reduce power imbalance between research worker and respondents, and attempts to change social inequality between men and women are the main characteristics of feminist research.

**Field editing of data:** Field editing of data consists in the review of the reporting forms by the investigator to complete or translating or rewriting, what the latter has written in abbreviated and/or in illegible form at the time of recording the respondents' responses.

**Field Report:** The field report contains the result of on-site field activity, evaluation of some on-going activities.

**Fifth generation computer:** The fifth generation computer is presently in the developing stage, may use the new switch (such as the High Electron Mobility Transistor) instead of the fourth one.

**Firmware:** Firmware is that software which is incorporated by the manufacturer into the electronic circuitry of the computer.

**First generation computer:** The first generation computer contained 18000 small bottles-sized valves which constituted its central processing unit (CPU).

**Fishbowl draws:** This is one of the methods of selecting a random sample and is useful particularly when 'N' is not very large. It entails writing each element number on a small slip of paper, folded and put into a bowl, shuffling thoroughly, and then taking one out till the required sample size is obtained.

**Fisher-Irwin Test:** Fisher-Irwin test is a distribution-free test employed in testing a hypothesis concerning no difference between two sets of data.

**Focus group:** The focus group is a form of strategy in qualitative research in which attitudes, opinions or perceptions towards an issue, product, service or programme are explored through a free and open discussion between members of a group and the research worker. The focus group is a facilitated group discussion in which a research worker raises issues or asks questions that stimulate discussion among members of the group. Issues, questions and different perspectives on them and any significant points arising during these discussions provide data to draw conclusions and inferences. It is like a collectively interviewing group of respondents.

**Focused interview:** Focused interview is described as to focus attention on the given experience of the respondent and its effects. The interviewer has the freedom to decide the manner and sequence in which the questions would be asked and has also the freedom to explore reasons and motives. The main task of the interviewer is to confine the respondent to a discussion of issues with which he seeks conversion. Such interviews are used normally in the development of hypotheses and constitute a major type of unstructured interviews.

**Footnote:** Footnote, given at the bottom of the same page, is meant for cross referencing, citing of authorities and sources, acknowledging and classifying, clarifying or explaining of a point of view but it is neither an end nor a means of the display of scholarship. Footnotes is always either numbered or symbolized.

**Formal experimental designs:** Formal experimental designs offer relatively more control and use precise statistical procedures for analysis.

**Formalized research:** Formalized research is that with substantial structure and with specific hypotheses to be tested.

**Formation of an index**: The formation of an index is combining the various indicators into an index.

**Formulative research studies:** *See exploratory research studies.*

**Fourth generation computers:** The fourth generation computers owe their birth to the advent of microprocessors. Because of the use of microprocessor as CPU on this computer device has enabled the development of microcomputers, personal computers, portable computers and the like.

**Frame of analysis:** The proposed plan of the way to analyze the data, the way to analyze the data to operate the major concepts and what statistical procedures are planned to use, these all form the parts of the frame of analysis.

**F-ratio:** F-ratio may be worked out as the ratio between MS between and MS within.

**Freedom from bias:** Freedom from bias is attained when the criterion gives each subject an equal opportunity to score well.

**Frequency distribution:** The frequency distribution is a statistical procedure in quantitative research that can be applied to any variable that is measured at any one of the four measurement scales. It groups respondents into the subcategories in which a variable has been measured or coded.

**Frequency polygon:** The frequency polygon is very similar to a histogram. A frequency polygon is drawn by joining the midpoint of each rectangle at a height commensurate with the frequency of that interval.

**F-test:** The F-test is based on the assumption of normality. F-test is based on F-distribution and is used to compare the variance of the two-independent samples.

**Fundamental research:** A research which is mainly concerned with generalization and with the formulation of a theory is called as fundamental research.

# G

**General article:** A general or popular article is like a short review paper based on an area of public interest. The author can incorporate his own idea on the particular topic in this article.

**Generation Computer:** The third generation computer followed the invention of the integrated circuit (IC). Such machines, with their CPU and main memory store made of IC chips.

**Geometric mean:** Geometric mean is defined as the $n^{th}$ root of the product of the values of n times in a given data series. It is also useful under certain conditions.

**Good design:** A good design is frequently characterized by adjectives like efficient, appropriate, economical, and flexible and so on. Generally, the design which minimizes bias and maximizes the reliability of the data collected and analyzed is considered a good design.

**Graphic rating scale:** The graphic rating scale is quite simple and is commonly used in practice. Under it the various points are usually put along the line to form a continuum and the rater indicates his rating by simply making a mark (such as u) at the appropriate point on a line that runs from one extreme to the other.

**Group interview:** A group interview is both a method of data collection and a qualitative study design. The interaction is between the research worker and the group with the aim of collecting information from the group collectively rather than individually from members.

**Guttmann scale:** The Guttmann scale is one of the three attitudinal scales and is devised in such a way that the statements or items reflecting attitude are arranged in perfect cumulative order. Arranging statements or items to have a cumulative relation between them is the most difficult aspect of constructing this scale.

# H

**Halo effect:** When making an observation, some observers may be influenced to rate an individual on one aspect of the interaction by the way he was rated on another. This is similar to something that can happen in teaching when a teacher's assessment of the performance of a student in one subject may influence his rating of that student's performance in another. This type of effect is known as the halo effect.

**Hardware:** Hardware is all the physical components (such as CPU, Input-output devices, storage devices, etc.) of the computer.

**Harmonic mean:** Harmonic mean is defined as the reciprocal of the average of reciprocals of the values of items of a data series.

**Hawthorne effect:** When individuals or groups become aware that they are being observed, they may change their behaviour. Depending upon the situation, this change could be positive or negative–it may increase or decrease, for example, their productivity–and may occur for a number of reasons. When a change in the behaviour of persons or groups is attributed to their being observed, it is known as the Hawthorne effect.

**Histogram:** A histogram is a graphic presentation of analyzing data presented in the form of a series of rectangles drawn next to each other without any space between them, each representing the frequency of a category or subcategory.

**Historical research:** Historical research is the research that utilizes historical sources like documents, etc. to study events or ideas of the past at any remote point of time.

**Holistic research:** Holistic research is more a philosophy than a study design. The design is based upon the philosophy that as a multiplicity of factors interacts in our lives, we cannot understand a phenomenon from one or two perspectives only. To understand a situation or phenomenon we need to look at it in its totality or entirety; that is, holistically from every perspective. A research study done with this philosophical perspective in mind is called holistic research.

**Holtzman Inkblot Test (HIT):** The Holtzman Inkblot Test (HIT) is a modification of the Rorschach Test. It contains more inkblot cards which are based on colour, movement, shading and other factors involved in inkblot perception. The respondents are asked to describe what they perceive in such

**Hypothesis of a difference:** A hypothesis in which a research worker stipulates that there will be a difference but does not specify its magnitude is called a hypothesis of a difference.

**Hypothesis of association:** When a research worker has sufficient knowledge about a situation or phenomenon and is in a position to stipulate the extent of the relationship between two variables and formulate a hunch that reflects the magnitude of the relationship, such a type of hypothesis formulation is known as hypothesis of association.

**Hypothesis of point-prevalence:** There are times when a research worker has enough knowledge about a phenomenon that he is studying and is confident about speculating almost the exact prevalence of the situation or the outcome in quantitative units. This type of hypothesis is known as a hypothesis of point-prevalence.

**Hypothesis:** A hypothesis can be explained as a proposition or a set of proposition set forth as an elucidation for the incidence of some specified group of phenomena either asserted simply as a provisional conjecture to guide some investigation or accepted as highly probable in the light of established facts.

**Hypothesis:** A hypothesis is an assumption, hunch, assertion, suspicion or an idea about a phenomenon, relationship or situation, the reality or the truth of which research worker is not known and he sets up his study to find this truth. A research worker refers to these assumptions, assertions, statements or hunches as hypotheses and it become the basis of an inquiry. In most studies the hypothesis will be based either upon previous studies or observations.

**Hypothesis-testing research studies:** *See Experimental research studies.*

**Hypothesis-testing research:** When the purpose of research is to test a research hypothesis, it is termed as hypothesis-testing research. It can be of the experimental design or of the non-experimental design.

# I

**IC:** Integrated circuit (IC) is a complete electronic circuit fabricated on a single piece of pure silicon.

**Illuminative evaluation:** The primary concern of illuminative or holistic evaluation is a description and interpretation rather than measurement and prediction of the totality of a phenomenon. It fits with the social–anthropological paradigm. The aim is to study a programme in all its aspects: how it operates, how it is influenced by various contexts, how it is applied, how those directly involved view its strengths and weaknesses, and what the experiences are of those who are affected by it. In summary, it tries to illuminate an array of questions and issues relating to the contents, and processes, and procedures that give both desirable and undesirable results.

**Impact assessment evaluation:** Impact or outcome assessment evaluation is one of the most widely practiced evaluations. It is used to assess what changes can be attributed to the introduction of a particular intervention, programme or policy. It establishes causality between an intervention and its impact, and estimates the magnitude of this change(s).

**Independent variable:** When examining causality in a study, there are four sets of variables that can operate. One of them is a variable that is responsible for bringing about change. This variable, which is the cause of the changes in a phenomenon, is called an independent variable. In the study of causality, the independent variable is the cause variable which is responsible for bringing about change in a phenomenon.

**In-depth interviewing:** In-depth interviewing is an extremely useful method of data collection that provides complete freedom in terms of content and structure. The research worker is free to order these in whatever sequence he wishes, keeping in mind the context. The research worker has complete freedom in terms of what questions he asks of your respondents, the wording you use and the way you explain them to his respondents. He usually formulates questions and raises issues on the spur of the moment, depending upon what occurs to him in the context of the discussion.

**Indicators:** An image, perception or concept is sometimes incapable of direct measurement. In such situations a concept is 'measured' through other means which are logically 'reflective' of the concept. These logical reflectors are called indicators.

**Indirect interviewing techniques:** *See Projective techniques.*

**Inferential approach to research**: It is to form a database from which to infer characteristics or relationships of population. This usually means survey research where a sample of population is studied (questioned or observed) to determine its characteristics, and it is then inferred that the population has the same characteristics.

**Informal experimental design:** Informal experimental design is that typically utilizes a less advanced type of analysis based on differences in magnitudes.

**Informed consent:** Informed consent implies that respondents are made adequately and accurately aware of the type of information wanted from them, why the information is being sought, what purpose it will be put to, how they are expected to participate in the study, and how it will directly or indirectly affect them. It is important that the consent should also be voluntary and without any kind of pressure. The consent given by respondents after being adequately and accurately made aware of or informed about all aspects of a study is called informed consent.

**Integrated circuit:** *See IC.*

**Internal validity:** The internal validity of a research design is its ability to measure what it aims to measure.

**Interpretability:** Interpretability consideration is particularly important when persons, except the designers of the test are to interpret the results. The measuring tool, in order to be interpretable, must be supplemented by (i) detailed tools for administering the test; (ii) scoring keys; (iii) evidence about the reliability and (iv) guide for utilizing the test and for interpreting results.

**Interpretation:** Interpretation refers to the task of drawing inferences from collecting facts after an analytical and/or experimental study.

**Interrupted time-series design:** In this design a group of people can be studied before and after the introduction of an intervention. It is like the before-and-after design, except that it needs multiple data collections at different time intervals to constitute an aggregated before-and-after picture. The design is based upon the assumption that one set of data is not sufficient to establish, with a reasonable degree of certainty and accuracy, the before-and-after situations.

**Interval scale:** The interval scale is one of the measurement scales where the scale is divided into a number of intervals or units. An interval scale has all the characteristics of an ordinal scale. In addition, it has a unit of measurement that enables individuals or responses to be placed at equally spaced intervals in relation to the spread of the scale. This scale has a starting and a terminating point and is divided into equally spaced intervals. The starting and terminating points and the number of intervals between them are arbitrary and vary from scale to scale as it does not have a fixed zero point. In interval scale, the intervals are adjusted in terms of some rule that has been established as a basis for making the units equal. The units are equal only in so far as the assumptions are acceptable on which the rule is based. Interval scales can have an arbitrary zero, but it is not possible to determine an absolute zero or the unique origin.

**Intervening variables:** An intervening variable is such type of factor which influences the examined phenomenon but cannot be seen and measured or manipulated, Its impact must be inferred from the effects of the Independent and moderator variables on the observed phenomena. These types of variables link the independent and dependent variables. In certain situations the relationship between an independent and a dependent variable does not eventuate till the intervention of another variable the intervening variable. The cause variable will have the assumed effect only in the presence of an intervening variable.

**Intervention development evaluation process:** This is a cyclical process of continuous assessment of needs, intervention and evaluation. You make an assessment of the needs of a group or community, develop intervention strategies to meet these needs, implement the interventions and then evaluate them for making informed decisions to incorporate changes to enhance their relevance, efficiency and effectiveness. Reassess the needs and follow the same process for intervention–development–evaluation.

**Interview guide:** A list of issues, topics, or discussion points that is necessary to cover in an in-depth interview is called an interview guide. These points are not questions. It is basically a list to remind an interviewer of the areas to be covered in an interview.

**Interview schedule:** An interview schedule is a written list of questions, open ended or closed, prepared for use by an interviewer in a person-to-person interaction (this may be face to face, by telephone or by other electronic media). An interview schedule is a research tool or instrument for collecting data, whereas interviewing is a method of data collection.

**Interview:** The interview is a technique for gathering data, including presentation of oral-verbal stimuli and reply in terms of oral-verbal responses.

**Interviewer bias:** Interviewer bias means to the extent to which an answer has altered in meaning by some action or attitude on the part of the interviewer.

**Interviewing**: The method of interviewing is one of the commonly used methods of data collection. Any person-to-person interaction, either face to face or otherwise, between two or more individuals with a specific purpose in mind is called an interview. It involves asking questions of respondents and recording their answers. Interviewing spans a wide spectrum in terms of its structure. On the one hand, it could be highly structured and, on the other, extremely flexible, and in between it could acquire any form.

**Item analysis approach of scaling:** Item analysis approach of scaling it a number of individual items is developing into a test which is given to a group of respondents. Cumulative scales are preferred based on their conforming to some ranking of items with ascending and descending discriminating power.

**Itemized rating scale:** The itemized rating scale (also known as numerical scale) presents a series of statements from which a respondent selects one as best reflecting his evaluation. The statements are ordered increasingly in terms of more or less of some property.

# J

**Judgment sampling:** Judgment sampling is utilized frequently during qualitative research to develop hypotheses rather than to generalize to larger populations. The research worker's decision is used for selecting items which are considered as representative of the population in judgment sampling.

**Judgmental sampling:** *See deliberate sampling.*

# K

**Karl Pearson's coefficient:** Karl Pearson's coefficient of correlation (also simple correlation) is the most widely used methods of measuring the degree of relationship between two variables. This coefficient assumes that there is a linear relationship between the two variables; the two variables are causally related which means that one of the variables is independent and the other one is dependent; and a large number of independent causes are operating in both variables so as to produce a normal distribution.

**Kendall's Coefficient of Concordance:** Kendall's coefficient of concordance, symbolized by W, is a non-parametric estimation of the relationship and works out the degree of association among several (k) sets of ranking of 'N' objects or individuals.

**Kurtosis:** Kurtosis is the measure of flat-toppedness of a curve of a data series. It is the humpedness of the curve and points to the nature of distribution of items in the middle of a series.

# L

**Latent root:** *See Eigen value.*

**Latent Structure Analysis:** Latent Structure Analysis shares both of the objectives of factor analysis to extract latent factors as well as an express relationship of observed (manifest) variables with the factors as their indicators and to classify a population of respondents into guanine types. This analysis is suitable if the variables involved in the study do not possess dependency relationship and happen to be non-metric.

**Latin square design (LSD):** Latin square design is an experimental design very frequently used in agricultural research. LSD is so allocated among the plots that no treatment arises more than once in any one row or column. The two blocking factors may be represented through rows and columns (one through rows and the other through columns). In such a design LSD the treatments are so allocated among the plots that no treatment occurs, more than once in any one row or any one column. The ANOVA technique in case of LSD remains more or less similar to that as in two-way design, excepting the fact that the variance is split into four parts, i.e. variance between columns; the variance between rows; variation between varieties; and residual variance.

**Leading question:** A leading question is one which, by its contents, structure or wording, leads a respondent to answer in a certain direction.

**Leptokurtic:** If in a data series a curve is relatively more peaked than the normal curve, it is called Leptokurtic.

**Likert scale:** The Likert scale, also known as the summated rating scale, is one of the attitudinal scales designed to measure attitudes. This scale is based upon the assumption that each statement or item on the scale has equal attitudinal 'value', 'importance' or 'weight' in terms of reflecting an attitude towards the issue in question. Comparatively it is the easiest to construct.

**Literature review:** A compilation of all literature available on a particular topic over a certain period is termed as literature review or review of literature.

**Literature review:** This is the process of searching the existing literature relating to research problem to develop theoretical and conceptual frameworks for study and to integrate research findings with what the literature says about them. It places studied in perspective to what others have investigated about the present issue. In addition the process helps to improve research methodology.

**Logical treatment in report writing:** Logical treatment often consists in developing the material from the simplest possible to the most complex structures.

**Long term trend:** *See Secular trend.*

**Longitudinal study:** In longitudinal studies, the study population is visiting a number of times at regular intervals, usually over a long period, to collect the required information. These intervals are not fixed so their length may vary from study to study. Intervals might be as short as a week or longer than a year. Irrespective of the size of the interval, the information gathered each time is identical.

**Louis Guttmann's scalogram analysis**: *See cumulative scale.*

**LSD**: *See Latin-square design.*

# M

**Matching:** The matching is a technique that is used to form two groups of subjects to set up an experiment–control study to test the effectiveness of a treatment. From a pool of subjects, two subjects with identical predetermined attributes, characteristics or conditions are matched and then randomly placed in either the experimental or control group. The process is called matching. The matching continues for the rest of the pool. The two groups, thus formed through the matching process are supposed to be comparable thus ensuring uniform impact of different sets of variables on the subjects.

**Mathematical basis of factor analysis:** The mathematical basis of factor analysis is concerned with a data matrix, also termed as score matrix, and symbolized as S.

**Maturation effect:** If the study population is very young and if there is a significant time lapse between the before-and-after sets of data collection, the study population may simply change because it is growing older. This is particularly true when young children are being studied. The effect of this maturation, if it is significantly correlated with the dependent variable, is reflected in the 'after' observation and is known as the maturation effect.

**Maximum Likelihood Method of Factor Analysis:** The Maximum Likelihood (ML) method consists in obtaining sets of factor loadings successively in such a way that each, in turn, explains as much as possible population correlation matrix as estimated from the sample correlation matrix.

**Maxmincon principle of variety:** When studying the causality between two variables there are three sets of variations that impact upon the dependent variable. Since the aim of a research worker is to determine the change that can be attributed to the independent variable, he needs to design the study to ensure that the independent variable has the maximum opportunity to have its full impact on the dependent variable, while the effects that are attributed to extraneous and chance variables are minimized. Setting up a study to achieve the above is known as adhering to the maxmincon principle of variety.

**McNemer Test:** McNemer test is one of the important non-parametric tests frequently used if the data is nominal and related to two related samples.

**Mean deviation:** Mean deviation is the average of the difference of the values of items from some average of the data series.

**Mean:** Mean may be defined as the result of dividing the total of the values of various given items in a series by the total number of items.

**Measurement:** Measurement is a procedure of mapping aspects of a domain onto additional aspects of a range according to some rule of correspondence. In measuring, some form of scale in the range is devised followed by transforming or mapping the properties of objects from the domain onto this scale.

**Median:** Median is the value of the middle item of a data series if it is arranged in ascending or descending order of magnitude or array.

**Memory chips:** The memory chips are the ICs from the secondary memory or storage of the computer. They can hold data and instructions.

**Mesokurtic:** A normal curve is Mesokurtic because it is kurtic in the center.

**ML Method of Factor Analysis:** See Maximum Likelihood Method of Factor Analysis.

**Mode:** Mode is the most common or frequently occurring value in a data series.

**Moderator variable:** The moderator variable may be defined as a such type of factor which can be measured, manipulated or selected by the research worker to discover whether it modifies the relationship of independent variable to an observed phenomena.

**Motivation Research:** Motivation research is a research aimed to discover the underlying motives and desires, using in depth interviews for the purpose.

**Multidimensional scaling:** Multidimensional scaling measures that an object might be described best by using the concept of an attribute space of 'n' dimensions, rather than a single-dimension continuum. Multidimensional scaling allows a research worker for measuring an item in more than one dimension at a time. The basic assumption in this scaling is that people perceive a set of objects are more or less similar to one another on a number of dimensions which is usually unrelated to one another, instead of only one. Multidimensional scaling can be characterized as a set of procedures for portraying perceptual or affective dimensions of substantive interest. It "provides a useful methodology for portraying subjective judgments of diverse kinds.

**Multifactor-factorial design:** *See* complex factorial design.

**Multiple Correlations:** When there are two or more than two independent variables, the correlation analysis concerning relationship is known as multiple correlations.

**Multiple discriminant analysis:** The discriminant analysis technique is used to classify individuals or objects into one of two or more mutually exclusive and exhaustive groups on the basis of a set of independent variables. Discriminant analysis necessitates interval independent variables and a nominal dependent variable. This type of analysis is appropriate in the case when the research worker has a single dependent variable that cannot be measured, but can be classified into two or more groups on the basis of some other attributes. The object of this type of analysis is to predict an entity's possibility of belonging to a particular group based on several predictor variables.

**Multiple regression analysis:** The multiple regression analysis is adopted in the case when the research worker has one dependent variable which is presumed to be a function of two or more independent variables. The objective of this process is to make a prediction about the dependent variable based on its covariance with all of the other concerned independent variables.

**Multiple regressions:** In multiple regressions a linear composite of explanatory variables is worked out in such a way that it has a maximum correlation with a criterion variable. The technique is suitable when the research worker has a single, metric criterion variable. In case if there are two or more than two independent variables, the equation describing such relationship is called as the multiple regression equation.

**Multi-stage random sampling:** Multi-stage random sampling is meant for big inquiries extending to a considerably large geographical area. Under this method of sampling the first stage may be to select large primary sampling units. Again the same is divided into sub areas. The technique of random-sampling is applied at all the stages so the sampling procedure is described as multi-stage random sampling.

**Multivariate analysis of variance:** Multivariate analysis of variance is just an extension of bivariate analysis of variance where the ratio of among-groups variance to within-groups variance is calculated on a set of variables rather than a single variable. This technique is considered best fit when several metric dependent variables are occupied in a research study along with many non-metric explanatory variables.

**Multivariate analysis of variance:** The multivariate analysis of variance analysis is an extension of two way ANOVA. The ratio of among group variance to within group variance is worked out on a set of variables.

**Multivariate analysis:** Multivariate analysis may be defined as "all statistical methods which simultaneously analyze more than two variables on a sample of observations.

**Multivariate population:** Multivariate population is the population consisting of measurement of more than two variables. In other words If for every measurement of a variable, X, has a corresponding value of the third variable, Z, or the forth variable, W, and so on, the resulting pairs of values are called a multivariate population.

**Multivariate techniques:** All statistical techniques, analyzing more than two variables on a sample of observations are known as multivariate techniques.

# N

**Narratives:** The narrative technique of gathering information has even less structure than the focus group. Narratives have almost no predetermined contents except that the research worker seeks to hear the personal experience of a person with an incident or happening in his life. Essentially, the person tells his story about an incident or situation and the research worker, listen passively, occasionally encouraging the respondent.

**News items:** News items appearing in the daily news papers are also forms of report writing. It represents firsthand, on-the-scene accounts of the events described or compilations of interviews. In such reports the first paragraph usually contains the most important information in detail and the succeeding paragraphs contain material which has progressively less and less importance.

**Newsletters:** The purpose of a newsletter is to communicate quickly facts that are of interest to its readers. Thus the content of any contribution is basically factual, with little highlighting on explanation or methodology.

**Nomenclature:** A system of names for things and scientific terminology, etc. or systematic naming is called nomenclature.

**Nominal data:** Nominal data are numerical in name only, because they do not share any of the properties of the numbers dealt in ordinary arithmetic.

**Nominal scale:** Nominal scale is simply a system of allotment of number signs to events in order to label them. This type of scale is one of the ways of measuring a variable. It enables the classification of individuals, objects or responses based on a common or shared property or characteristic. These respondents or objects or responses are divided into a number of subgroups in such a way that each member of the subgroup has the common characteristic.

**Non-continuous variables:** *See discrete variables.*

**Non-directive interview:** A non-directive interview is the interview if the interviewer's function is basically to encourage the respondent to discuss about the given topic with an absolute minimum of direct questioning. The interviewer regularly acts as a catalyst to a comprehensive expression of the respondents' emotions and convictions and of the edge of reference in which such feelings and convictions take on personal significance.

**Non-experimental studies:** There are times when, in studying causality, a research worker observes an outcome and wishes to investigate its causation. From the outcomes the research worker starts linking causes with them. Such studies are called non-experimental studies. In a non-experimental study the cause variable is neither introduces nor controlled or manipulated.

**Non-metric approach:** Non-metric approach inquires about a representation of points in a space of minimum dimensionality such that the rank order of the inter-point distances in the solution space maximally corresponds to that of the data. This is achieved by requiring only that the distances in the solution are monotone with the input data.

**Non-parametric tests:** Non-parametric tests are the tests of hypothesis that do not depend on any assumption about the parameters of the parent population.

**Non-participant observation:** The non-participant observation is the observation when the observer observes as a detached emissary without any attempt on his part to experience through participation what others feel.

**Non-participant observation:** When A research worker, does not get involved in the activities of the group, but remain a passive observer, watching and listening to its activities and interactions and drawing conclusions from them, this is called non-participant observation.

**Non-probability sampling designs:** Non-probability sampling designs do not follow the theory of probability in the selection of elements of the sampling population. Non-probability sampling designs are used when the number of elements in a population is either unknown or cannot be individually identified. In such situations the selection of elements is dependent upon other considerations. Non-probability sampling designs are commonly used in both quantitative and qualitative research.

**Non-probability sampling:** *See deliberate sampling.*

**Non-sampling errors:** Non-sampling errors are procedural error which may creep in during the process of collecting actual data and such errors are expected to be occurred in all the experiments and there is no statistical method to measure non-sampling error.

**Non-specific hypothesis:** *See Composite hypothesis.*

Non-structured questionnaire: The unstructured questionnaire is one in which questions and answers are not specified and comments are not compulsory to be in the respondent's own words. This is not definite, concrete and may contain simultaneous questions. The questions (may be subjective or descriptive) are presented with flexibility in terms or wording and the order to all respondents.

**Non-technical writing:** Non-technical writing deals with the literature which concerns to sell to the public for entertainment or advertisement.

**Normal curve:** If the distribution of item in a data series happens to be perfectly symmetrical, having mean=median=Z score.

**Null hypothesis:** The null hypothesis is that the population mean is equal to the hypothesized mean and symbolically it is expressed as $H_0$: $\mu=\mu_{H0}$. When a hypothesis is constructed stipulating that there is no difference between the two situations, groups, outcomes, or the prevalence of a condition or phenomenon, this is called a null hypothesis and is usually written as *H*0.

**Numerical scale:** *See Itemized rating scale.*

**Paired comparison:** Under Paired comparison the respondent can express his attitude by making a choice between two objects.

# O

**Objective- oriented evaluation:** The objective- oriented evaluation is the evaluation when it is designed to ascertain whether or not a programme or a service is achieving its objectives or goals.

**Observation Method:** The observation method is the most frequently used techniques as a research method. In a way the research worker observes the items around, but this kind of observation is not a scientific observation. The observation has become a scientific tool and the technique of data collection for the research worker, only in the case when it serves a formulated research objective, is systematically designed and recorded and is subjected to checks and controls on validity and reliability.

**Observation:** Observation is one of the methods for collecting primary data. It is a purposeful, systematic and selective way of watching and listening to an interaction or phenomenon as it takes place. Though dominantly used in qualitative research, it is also used in quantitative research.

**Observational design:** The observational design relates to the conditions under which the observations are to be made.

**One sample runs Test:** There are lots of applications in which it is very much difficult to decide whether the sample used is a random one or not. One sample runs test is a test tooled to judge the randomness of a sample on the basis of the order in which the observations are taken.

**One sample sign test:** The one sample sign test is a very simple, non-parametric test applicable when a continuous symmetrical population in which case the probability of getting a sample value less than the mean is ½ and the probability of getting a sample value greater than the mean is also ½, is sampled.

**One-tailed tests:** If the population mean is either lower or higher than some hypothesized value the test is called one tailed test. Symbolically, the two tailed is presented as $H_0$: $\mu=\mu_{H0}$ and $\mu<\mu_{H0}$.

**One-way ANOVA:** Under the one-way ANOVA, only one factor is considered and then the reason for said factor to be important is observed.

**Open-access journals:** The concept of open access via the internet has become more common now a day. Open-access journals are available to anyone who has access to the internet subjected too much discussion.

**Open-ended questions:** In an open-ended question the possible responses are not given. In the case of a questionnaire, a respondent writes down the answers in his/her words, whereas in the case of an interview schedule the investigator records the answers either verbatim or in a summary describing a respondent's answer.

**Operating characteristic function:** Operating characteristic function shows the conditional probability of accepting $H_0$ for all values of population parameter(s) for a given sample size.

**Operating software:** Operating software or system software is that program which tells the computer how to function. It is also known as operating software and is normally supplied by the computer manufacturer.

**Operational definition:** When used concepts are defined either in the research problem or in the study population in a measurable form, they are called working or operational definitions. It is important to understand that the working definitions, that are developed, are only for the purpose of the study.

**Operational design:** The operational design concerns with the research methodology by which the procedures specified in sampling and statistical and observational designs can be performed.

**Opinionnaire:** An information form that attempts to measure the attitude or belief of an individual is known as opinionnaire.

**Oral history:** Oral history is more a method of data collection than a study design, however, in qualitative research, it has become an approach to study a historical event or episode that took place in the past or for gaining information about a culture, custom or story that has been passed on from generation to generation. It is a picture of something in someone's own words. Oral histories, like narratives, involve the use of both passive and active listening. Oral histories, however, are more commonly used for learning about cultural, social or historical events, whereas narratives are more about a person's own experience.

**Oral presentation:** An oral presentation of the results of the study is considered effective, particularly in cases where policy recommendations are specified by project results. The merit of this approach mendacity in the fact that it offers an opportunity for give-and-take decisions which generally lead to a better understanding of the findings and their implications. This type of presentation of research results is a major method of communicating the results of a research endeavour. This pattern is commonly adopted in institutional seminars, international conferences, workshops and training courses.

**Ordinal data:** In those situations when the data can only be set up inequalities, data is referred to be an ordinal data.

**Ordinal scale:** An ordinal scale has all the properties of a nominal scale plus one of its own. Besides categorizing individuals, objects, responses or a property into subgroups on the basis of a common characteristic, it ranks the subgroups in a certain order, arranged in either ascending or descending order according to the extent that a subcategory reflects the magnitude of variation in the variable. This type of scale places, events in order, but there is no effort to make the intervals of the scale identical in terms of some rule. Rank orders symbolize ordinal scales and are frequently exercised in research relating to qualitative phenomena.

**Outcome evaluation:** The focus of an outcome evaluation is to find out the effects, impacts, changes or outcomes that the programme has produced in the target population.

**Outlines:** Outlines are the framework constructing foundation to long written works. Those are an aid to the logical organization of material and a reminder of the stressed points in the report.

# P

**P.C. Method of factor analysis:** Principal-component's method or simply P.C. Method of factor analysis, was developed by Hotelling. This \method searches to maximize the sum of squared loadings of each factor extracted. Accordingly P.C. factor determines more variance than would the loadings obtained from any other method of factoring.

**Paired t-test:** Paired t-test is a method to test for comparing two related samples, involving small values of 'n' that does not require the variances of the two populations and supposed to be equal, but the assumption that the two populations are normal must continue to apply.

**Panel studies:** Panel studies are prospective in nature and are designed to collect information from the same respondents over a period of time. The selected group of individuals becomes a panel that provides the required information. In a panel study the period of data collection can range from once only to repeat data collections over a long period.

**Pantry audit technique:** The pantry audit technique is used for estimating consumption of the basket of goods at the consumer level. In this type of audit, the investigator gathers an inventory of types, quantities and prices of commodities consumed. Consequently, in pantry audit data are recorded from the examination of the consumer's pantry.

**Parameter:** A parameter is working out of certain statistical measures from the population or universe to describe its characteristics.

**Parametric tests:** Parametric tests usually assume certain properties of the parent population from which samples are drawn.

**Partial correlation:** The partial correlation is a type of correlation to measure the relation between a dependent variable and a particular independent variable by holding all other variables constant.

**Participant observation**: Participant observation is the observation when the observer observes by making himself a member of the group that he is observing to experience what the members of the group experience. This kind of observation is when a research worker, participate in the activities of the group being observed in the same manner as its members, with or without knowing that those are being observed. Participant observation is principally used in qualitative research and is usually done by developing a close interaction with members of a group or 'living' in with the situation which is being studied.

**Participatory research:** Both participatory research and collaborative inquiry are not studying designs per se, but signify a philosophical perspective that advocates an active involvement of research participants in the research process. Participatory research is based upon the principle of minimizing the 'gap' between the research worker and the research participants. The most important feature is the involvement and participation of the community or research participants in the research process to make the research findings more relevant to their needs.

**Path analysis:** The term 'path analysis' first introduced by Wright in 1934, based on a series of multiple regression analyses with the added assumption of a causal relationship between independent and dependent variables.

**Periodic Report:** This type of report is submitted at regular intervals to the administration to provide the information on the progress or status of ongoing activities.

**Periodicals:** A periodical is defined as a publication issued in successive parts, usually at regular intervals, and as a rule, intended to be continued indefinitely.

**Personal interviews:** Personal interview is a strategy for gathering data which requires a person known as the interviewer to pose the inquiries generally in a face-to-face contact with the other person or persons. At the circumstances the interviewee may also pose certain questions and the interviewer responds to those, but generally the interviewer initiates the interview and gathers the information.

**Pie chart:** The pie chart is another way of representing data graphically. As there are 360 degrees in a circle, the full circle can be used to represent 100 per cent of the total population. The circle or pie is divided into sections in accordance with the magnitude of each subcategory comprising the total population. Hence, each slice of the pie is in proportion to the size of each subcategory of a frequency distribution.

**Pilot study:** *See feasibility study.*

**Placebo effect:** A patient's belief that he is receiving the treatment plays an important role in his recovery even though the treatment is fake or ineffective. The change occurs because a patient believes that he is receiving the treatment. This psychological effect that helps a patient to recover is known as the placebo effect.

**Placebo study:** A study that attempts to determine the extent of a placebo effect is called a placebo study. A placebo study is based upon a comparative study design that involves two or more groups, depending on whether or not it is needfull to have a control group to isolate the impact of extraneous variables or other treatment modalities to determine their relative effectiveness.

**Plagiarism:** Plagiarism is passing off others' words or ideas as one's own. It is considered as scientific misconduct.

**Platykurtic:** If in a data series, the curve is flatter than the normal curve, it is called Platykurtic.

**Play techniques:** In play techniques, subjects are asked to improvise or act out a situation in which they have been assigned various roles.

**Polytomous variable:** When a variable can be divided into more than two categories, for example religion (Christian, Muslim, Hindu), and attitudes (strongly favorable, favorable, uncertain, unfavorable, strongly unfavourable) etc., it is called a polytomous variable.

**Popular article:** *See general article.*

**Popular Report:** The popular report is one of the technical writing, which gives emphasis on simplicity and attractiveness. The simplification in popular report means clear writing, minimization of technical details and liberal use of charts, tables and diagrams. Attractive layout includes along with large print and division in many subheadings.

**Population mean:** From what is found out from the sample (sample statistics) it is made an estimate of the prevalence of these characteristics of the total study population. The estimates about the total study population made from sample statistics are called population parameters or the population mean.

**Population:** Population refers to the total of items about which information is desired.

**Precision:** The precision is the range within which the population mean will lie in reference to the reliability indicated at the confidence level as a level in terms of percentage of the estimate or as a numerical quantity.

**Predictive validity:** Predictive validity is judged by the degree to which an instrument can correctly forecast an outcome: the higher the correctness in the forecasts, the higher the predictive validity of the instrument. This kind of validity former refers to the usefulness of a test in predicting some future performance.

**Pre-test:** In quantitative research, pre-testing is a practice whereby something is tested that is developed before its actual use to ascertain the likely problems with it. Mostly, the pre-test is done on a research instrument or on a code book. The pre-test of a research instrument entails a critical examination of each question as to its clarity, understanding, wording and meaning as understood by potential respondents with a view to remove possible problems with the question. It ensures that a respondent's understanding of each question is in accordance with original intentions. The pre-test of an instrument is only done in structured studies. Pre-testing a code book entails actually coding a few questionnaires/interview schedules to identify any problems with the code book before coding the data.

**Primary data:** Information collected for the specific purpose of a study either by the research worker or by someone else is called primary data. This type of data are the data which are gathered anew and for the first time, and therefore happen to be unique in character.

**Primary Literature:** Primary literature consists of original research articles in scientific journals. It is highly objective as it based on observed data and will have been carefully reviewed to ensure that it conforms to rigorous scientific investigation principles by competent experts before being published.

**Primary sources:** Sources that provide primary data such as interviews, observations, and questionnaires are called primary sources.

**Principal-components Method of Factor Analysis:** *See P.C. Method of factor analysis.*

**Probability sampling:** *See Chance sampling.*

**Probability sampling:** When selecting a sample, if research worker adheres to the theory of probability, that is he selects the sample in such a way that each element in the study population has an equal and independent chance of selection in the sample, the process is called probability or random sampling. For a design to be called probability or random sampling, it is imperative that each element in the study population has an equal and independent chance of selection in the sample. Equal implies that the probability of selection of each element in the study population is the same. The concept of independence means that the choice of one element is not dependent upon the choice of another element in the sampling.

**Proceedings:** The proceeding contains the details of the papers presented at national or international seminar or talk or in a poster session during a particular conference.

**Process evaluation:** The main emphasis of process evaluation is on evaluating the manner in which a service or programme is being delivered in order to identify ways of enhancing the efficiency of the delivery system.

**Programme planning evaluation:** Before starting a large-scale programme it is desirable to investigate the extent and nature of the problem for which the programme is being developed. When an evaluation is undertaken with the purpose of investigating the nature and extent of the problem itself, it is called programme planning evaluation.

**Progress Report:** The progress report contains the updates of on-going activity as it is being carried out.

**Project proposals:** A project proposal represents the justification for a programme of work, with the aim to produce measurable outputs that will demonstrably reach a clearly defined objective.

**Projective techniques:** Projective techniques (or indirect interviewing techniques) for the collection of information are created to utilize projections of respondents for deriving about fundamental motives, urges, or intentions, which are to such an extent that the respondent either opposes to reveal them or is unable to figure out himself.

**Property of consistency:** Property of consistency is a property that an estimator should approach the value of the population parameter as the sample size becomes larger and larger.

**Property of efficiency:** Property of efficiency is a property that an estimator has relatively small variance, i.e. the most efficient estimator, among a group of unbiased estimators, is one which has the smallest variance. This is a property that an estimator is used as much as possible the information available from the sample.

**Property of unbiasedness:** Property of unbiasedness is a property that an estimator is on the average being equal to the value of the parameter being estimated.

**Proportionate stratified sampling:** In proportionate stratified sampling, the number of elements selected in the sample from each stratum is in relation to its proportion in the total population. A sample thus selected is called a proportionate stratified sample.

**Prospective studies:** Prospective studies refer to the likely prevalence of a phenomenon, situation, problem, attitude or outcome in the future. Such studies attempt to establish the outcome of an event or what is likely to happen. Experiments are usually classified as prospective studies because the research worker must wait for an intervention to register its effect on the study population.

**Pure research:** Pure research can be characterized as the improvement, examination, confirmation and refinement of research strategies, systems, procedures and instruments that frame the assortment of research technique.

**Purposive sampling:** *See deliberate sampling.*

# Q

**Q-type factor analyses:** In Q-type factor analysis, the correlations are calculated between pairs of respondents in place of pairs of variables.

**Qualitative approach to research:** Qualitative approach to research is concerned with subjective assessment of attitudes, opinions and behaviour. This type of approach to research generates findings either in non-quantitative form or in the form which are not concerned with the rigorous quantitative analysis. Commonly, the techniques of focus group interviews, projective techniques, and depth interviews are used.

**Qualitative research:** There are two broad approaches to inquire the facts or information: qualitative and quantitative or unstructured and structured approaches. Qualitative research is based upon the philosophy of empiricism, follows an unstructured, flexible and open approach to inquire, aims to describe than measure, believes in in-depth understanding and small samples, and explores perceptions and feelings than facts and figures.

**Qualitative research:** Qualitative research is concerned with qualitative phenomena relating to or involving quality or kind.

**Quantitative characteristics of data:** Quantitative characteristics of the data refer to a quantitative phenomenon which can be measured quantitatively or in the numeric.

**Quantitative research:** Quantitative research is based on the measurement of quantity or amount. It is applicable to phenomena that can be expressed in terms of quantity. Quantitative research is a second approach to inquire that is rooted in rationalism, follows a structured, rigid, predetermined methodology, believes in having a narrow focus, emphasizes greater sample size, aims to quantify the variation in a phenomenon, and tries to make generalizations to the total population.

**Quasi-experiments:** Studies which have the attributes of both experimental and non-experimental studies are called quasi- or semi-experiments. A part of the study could be experimental and the other non-experimental.

**Question:** Question is statement consisting at least questioning word and ends with a question mark (?).

**Question sequence:** Question sequence is an order of the questions in the questionnaire to make it effective and to ensure quality to the replies a research worker receives.

**Questionnaire:** A questionnaire is a written list of questions, the answers to which are recorded by respondents. In a questionnaire respondents read the questions, interpret what is expected and then write down the answers. The only difference between an interview schedule and a questionnaire is that in the former it is the interviewer who asks the questions (and, if necessary, explains them) and records the respondent's replies on an interview schedule, while in the latter replies are recorded by the respondents themselves.

**Questionnaire:** The questionnaire is a tool to gather information, which consists of a number of questions printed or typed in a definite order in a format or set of formats. The questionnaire is mailed to the respondents who are expected to read and understand the questions and write down the reply in the space meant for the purpose in the questionnaire itself. The respondents have to answer the questions on their own.

**Quizzes test:** The quizzes test is a type of projective techniques of interviewing for extracting information regarding the specific ability of the respondent indirectly. In this procedure questions are framed to test through them the memorizing and analytical ability of candidates.

**Quota sampling:** In quota sampling interviewers are simply given quota to be filled from different strata, the actual selection of items for sample being left to the interviewer's decision. The size of the quota for each stratum is generally proportionate to the size of that stratum in the same population. Quota sampling is thus an important form of non-probability sampling.

**Quota sampling:** The main consideration directing quota sampling is the research worker's ease of access to the sample population. In addition to convenience, a research worker is guided by some visible characteristic of interest, such as gender or race, of the study population. The sample is selected from a location convenient to a research worker, and whenever a person with this visible, relevant characteristic is seen, that person is asked to participate in the study. The process continues until to contact the required number of respondents (quota).

# R

**Random design:** In a random design, the study population groups as well as the experimental treatments are not predetermined, but randomly assigned to become controlling or experimental groups. Random assignment in experiments means that any individual or unit of the study, the population has an equal and independent chance of becoming a part of the experimental or control group or, in the case of multiple treatment modalities; any treatment has an equal and independent chance of being assigned to any of the population groups. It is important to note that the concept of randomization can be applied to any of the experimental designs.

**Random variable:** *See chance variable.*

**Randomization:** In experimental and comparative studies, it is often needful to study two or more groups of people. In forming these groups it is important that they are comparable with respect to the dependent variable and other variables that affect it so that the effects of independent and extraneous variables are uniform across groups. Randomization is a process that ensures that each and every sample in a group is given an equal and independent chance of being in any of the groups, thereby making groups comparable.

**Randomized block design (R.B.D.):** Randomized block design is an improvement over the C.R.D. In the R.B.D. the principle of local control can be applied along with the other two principles of experimental designs. In the R.B.D., samples are first divided into groups, known as blocks.

**Range:** Range is the simplest possible measure of dispersion and is defined as the difference between the values of the extreme items of a data series.

**Rank correlation:** *See Charles Spearman's coefficient.*

**Rank order:** Rank order is a method of comparative scaling, in which a respondent is asked to rank their choices. This method is easier and faster than other methods of paired comparisons.

**Ranking scales:** *See comparative scale.*

**Ratio scale:** A ratio scale has all the properties of nominal, ordinal and interval scales plus its own property; the zero point of a ratio scale is fixed, which means it has a fixed starting point. Therefore, it is an absolute scale.

As the difference between the intervals is always measured from a zero point, arithmetical operations can be performed on the scores. Ratio scales have an absolute or true zero of measurement. The term 'absolute zero' is not as precise as it was once believed to be.

**Reactive effect:** Sometimes the way a question is worded informs respondents of the existence or prevalence of something that the study is trying to find out about as an outcome of an intervention. This effect is known as reactive effect of the instrument.

**Recall error:** Such type of error that can be introduced in a response because of a respondent's inability to recall correctly its various aspects when replying.

**Reference list:** The reference list is the place where all the details are given what a reader needs to find the work being cited.

**References:** *See bibliography.*

**Reflective journal logs:** Basically, this is a method of data collection in qualitative research that entails keeping a log of the thoughts of a research worker whenever anything is noticed, talk to someone, participate in an activity or observe something that helps in understanding or adding to whatever is tried to find out about. This log becomes the basis of the research findings.

**Reflexive control design:** In experimental studies, to overcome the problem of comparability in different groups, sometimes research workers study only one population and treat data collected during the non-intervention period as representing a control group, and information collected after the introduction of the intervention as if it pertained to an experimental group. It is the periods of non-intervention and intervention that constitute control and experimental groups.

**Regression effect:** Sometimes people who place themselves in the extreme positions on a measurement scale of the pre-test stage may, for a number of reasons, shift towards the mean at the post-test stage. They might feel that they have been too negative or too positive at the pre-test stage. Therefore, the mere expression of the attitude in response to a questionnaire or interview has caused them to think about and alter their attitude towards the mean at the time of the post-test. This type of effect is known as the regression effect.

**Regression:** Regression is the determination of a statistical relationship between two or more variables.

**Relevance:** The relevance of a criterion is relevant if judged to be the proper measure.

**Reliability level:** *See Confidence level.*

**Reliability:** Reliability is the ability of a research instrument to provide similar results when used repeatedly under similar conditions. Reliability indicates accuracy, stability and predictability of a research instrument; the higher the reliability, the higher the accuracy; or the higher the accuracy of an instrument, the higher its reliability.

**Replicated cross-sectional design:** This type of study design is based upon the assumption that participants at different stages of a programme are similar in terms of their socioeconomic–demographic characteristics and the problem for which they are seeking intervention. Assessment of the effectiveness of an intervention is done by taking a sample of clients who are at different stages of the intervention. The difference in the dependent variable among clients at the intake and termination stage is considered to be the impact of the intervention.

**Repositories:** Technology has been developed to allow institutions to place their research content (theses, articles, working papers, conference proceedings, etc.) within a database that is then made available within the university's intranet, or to the world on the internet. These developments are called repositories.

**Research design:** A research design can be defined as a procedural plan that is adopted to answer questions validly, objectivity, accuracy and economy. A research design therefore answers questions that would determine the path proposed to take for research journey. Through a research design communicates the decisions regarding what study design proposed to use, to collect information from respondents, to analyze collected information and to communicate the findings. So the research design is an arrangement of suitable and best applicable conditions to collect and analyze the data in a manner that aims to combine relevance to the research purpose with economy in procedure.

**Research hypothesis:** When a prediction or a hypothesized relationship is to be tested by scientific methods, it is termed as research hypothesis. The research hypothesis is a predictive statement that relates an independent variable to a dependent variable.

**Research journals:** A periodical publishing scientific papers that communicate new and original information to other scientists.

**Research methodology:** Research methodology is a way to systematically solve the research problem. It may be defined as a science of studying how research is done scientifically. The research methodology is not only a matter to talk about the research methods, but also consider the logic behind the methods and explain why a particular method or technique is used and why another is not used so that the results may be evaluated either by the research worker himself or by others.

**Research methods:** Research methods may be defined as all those methods/ techniques that are used to conduct research. Research methods or techniques, thus, refer to the methods the research workers use in performing research operations.

**Research objectives:** Research objectives are specific statements of goals that are set out to be achieved at the end of the research journey.

**Research problem:** A research problem, in general, indicates to some complexity which a research worker experiences in the context of either a theoretical or practical situation and wants to obtain a solution for the same. Thus, any issue, problem or question that becomes the basis of inquiry is called a research problem. It is what is wanted to find out about during research endeavour.

**Research process:** Research process consists of a series of actions or steps necessary to effectively carry out research and the desired sequence of these steps.

**Research project proposals:** Research project proposals are, in essence, mini-research papers that have not reached the stage of actual implementation. Research proposals have a number of well defined elements that, if carefully put together, should result in successful funding for the project.

**Research questions:** Questions that is liked to find answers through the research, like 'What does it mean to have a child with ADHD in a family?' or 'What is the impact of immigration on family roles?' Research questions become the basis of research objectives. The main difference between research questions and research objectives is the way they are worked. Research questions take the form of questions, whereas research objectives are the statements of achievements expressed using action oriented words.

**Research report:** Research report is the major component and part and parcel of a research study. It is a presentation of research methods and results in a systematic manner. Writing of the report is considered as the last step in a research study.

**Research review paper:** The research review paper is a special type of scientific articles that, in many ways, is like an extended version of the discussion section of a research paper. An essential feature of a research review paper is that the reader is led to the frontiers of science in the area covered.

**Research techniques:** *See Research methods.*

**Research:** Research is one of the ways of finding answers to the professional and practice questions. However, it is characterized by the use of testing procedures and methods and an unbiased and objective attitude in the process of exploration. The term research in general refers to a search for knowledge or a scientific and systematic search for pertinent information on a particular topic. According to the Advanced Learner's Dictionary of Current English – it is a careful investigation or inquiry, especially through the search for new facts in any branch of knowledge. According to the Redman and Mory, it is a systematized effort to gain new knowledge.

**Retrospective prospective study:** A retrospective–prospective study focuses on past trends in a phenomenon and studies it into the future. A study where you measure the impact of an intervention without having a control group by 'constructing' a previous baseline from either respondents' recall or secondary sources, then introducing the intervention to study its effect, is considered a retrospective prospective study. In fact, most before-and-after studies, if carried out without having a control where the baseline is constructed from the same population before introducing the intervention will be classified as retrospective-prospective studies.

**Retrospective study:** A retrospective study investigates a phenomenon, situation, problem or issue that has happened in the past. Such studies are usually conducted either on the basis of the data available for that period or on the basis of respondents' recall of the situation.

**Review of literature:** *See literature review.*

**Reviews:** *See Research review paper.*

**Rorschach test**: The Rorschach test is a type of projective techniques of interviewing which consists of several cards having prints of inkblots. The design is symmetrical, but meaningless. The respondents are asked to describe what they perceive in such symmetrical inkblots.

**Rosenzweig test:** The Rosenzweig test is a type of projective techniques of interviewing which contains a cartoon format wherein words can be inserted in given balloons. The respondent is asked to put his own words in an empty balloon.

**Row percentages:** Row percentages are calculated from the total of all the subcategories of one variable that are displayed along a row in different columns.

**R-type factor analyses:** In R-type factor analysis, high correlations occur when respondents who score high on first variable also score high on second variable and respondents scoring low on first variable also score low on second variable.

# S

**Sample size:** The number of individuals from whom the required information is obtained, is called the sample size and is usually denoted by the letter *n*.

**Sample statistics:** Findings based on the information obtained from the respondents (sample) are called sample statistics.

**Sample:** A sample is a subgroup of the population, which is the focus of research inquiry and is selected in such a way that it represents the study population. A sample is composed of a few individuals from whom the required information is collected. It is done to save time, money and other resources.

**Sampling design:** A sampling design is referred to the technique or the procedure the research worker would adopt in selecting some sampling units from which inferences about the population is drawn. This type of design deals with the method of selecting items to be observed for the given study. The way the required sampling units are selected from a sampling population in identifying sample is called the sampling design or sampling strategy. There are many sampling strategies in both quantitative and qualitative research.

**Sampling distribution of proportion:** Sampling distribution of proportion describes the probability distribution of all the possible statistical measures of random samples of a given size taken from a population. This happens in the case of statistics of attributes.

**Sampling distribution of the mean:** Sampling distribution of mean describes the probability distribution of all the possible means of random samples of a given size taken from a population.

**Sampling element:** Anything that becomes the basis of selecting sample such as an individual, family, household, members of an organization, residents of an area, is called a sampling unit or element.

**Sampling error:** Sampling error can be measured for a given sample design and size. The measurement of sampling error is usually called the 'precision of the sampling plan'. The difference in the findings (sample statistics) that is due to the selection of elements in the sample is known as sampling error. This type of errors can be explained as such errors which arise on account of sampling and they generally happen to be random variations (in case of random sampling) in the sample estimates around the true population values.

**Sampling frame:** The sampling frame consists of a list of items from which the sample is to be drawn. When a research worker is in a position to identify all the elements of a study population, the list of all the elements is called a sampling frame.

**Sampling population:** The bigger the group, such as families living in an area, clients of an agency, residents of a community, members of a group, people belonging to an organization about whom it is needfull to find out about through research endeavour is called the sampling population or study population.

**Sampling strategy:** *See Sampling design.*

**Sampling theory:** Sampling theory can be defined as a study of relationships existing between a population and samples drawn from the population. This theory is applicable only to random samples.

**Sampling unit:** *See Sampling element.*

**Sampling with replacement:** When a sample is selected in such a way that each selected element in the sample is replaced back into the sampling population before selecting the next, this is called sampling with replacement. Theoretically, this is done to provide an equal chance of selection to each element so as to adhere to the theory of probability to ensure randomization of the sample. In case an element is selected again, it is discarded and the next one is selected. If the sampling population is fairly large, the probability of selecting the same element twice is fairly remote.

**Sampling without replacement:** When a sample is selected in such a way that an element, once selected to become a part of the sample, is not replaced back into the study population, this is called sampling without replacement.

**Sampling:** Sampling is the process of selecting a few respondents (a sample) from a bigger group (the sampling population) to become the basis for estimating the prevalence of information of interest. Sampling might be depicted as the selection of any part of any piece of a total or totality on the basis of which a derivation about the total or totality is made. In addition, this is a procedure of getting data regarding an entire population by assessing only a part of.

**Sandler's A-test:** Joseph Sandler has developed simplified alternate approach of Student's t-test relating to paired data, called as Sandler's A-test. The Sandler's 'A' value is the ratio between the sum of squares of the differences and the squares of the sum of the differences.

**Saturation point:** The concept of saturation point refers to the stage in data collection where a research worker, are discovering no or very little new information from the respondents. In qualitative research, saturation point is considered an indication of the adequacy of the sample size.

**Scale:** This is a method of measurement and/or classification of respondents on the basis of their responses to the questions asked. A scale could be continuous or categorical. It helps to classify a study population in subgroups or as a spread that is reflective of the scale.

**Scaling:** The scaling can be described as a procedure to assign numbers to various extents of opinion, attitude and other concepts. In other words, it can be illustrated as a process to assign numeric figures to various degrees of view, attitude and other concepts.

**Scatter-gram:** When it is required to show graphically how one variable changes in relation to a change in the other, a scatter-gram is extremely effective. For a scatter-gram, both the variables must be measured either on an interval or ratio scale and the data on both the variables needs to be available in absolute values for each observation. Data for both variables is taken in pairs and displayed as dots in relation to their values on both axes. The resulting graph is known as a scatter-gram.

**Schedules:** Schedule is a format containing a set of questions like questionnaires with little difference which lies in the fact that schedules are being filled in by the enumerators who are specially appointed for the purpose.

**Scientific poster:** A scientific poster is a single large page containing all the information to be communicated. In effect, a poster is an abridged form of a journal article, and therefore follows the IMRAD (Introduction, Materials, Results and Discussion) format.

**Scientific thinking:** The scientific thinking may be defined as an inductive-deductive mode of thinking or reasoning in which the uniformities of nature by appealing to experiences are explained. Induction moves forward from particular to the general whereas deduction backward movement.

**Scientific writing:** A technical writing is sometimes known as scientific writing when a topic is specifically scientific in nature.

**Score matrix**: *See Mathematical basis of factor analysis.*

**Second generation computer:** The second generation computer involved transistors which replaced the valve in all electronic devices and made them much smaller and more reliable.

**Secondary data:** The secondary data are those data which have already been collected by someone else and which have already been processed through the statistical analysis. Sometimes the information required is already available in other sources such as journals, previous reports, censuses and that the information can be extracted for the specific purpose of the research study. This type of data which already exists but needs to be extracted for the purpose of study is called secondary data.

**Secondary literature:** The secondary literature contains the literature, which survives from primary literature after the process of testing and scrutiny by the wider scientific community is over. It is referred in reputed scientific journals, in scientific articles, and in review articles.

**Secondary sources:** Sources that provide secondary data are called secondary sources. Sources such as books, journals, previous research studies, records of an agency, client or patient information already collected and routine service delivery records all form secondary sources.

**Secular trend:** Secular trend or long term trend is the time series showing the direction of the series in a long period of time.

**Semantic differential scale:** Semantic differential scale is an attempt to measure the psychological meaning of an object to an individual. Semantic differential scale is based on the presumption that an object can have different dimensions of connotative meanings which can be located in multidimensional property space in the context of this scale.

**Semi-experimental design:** A semi-experimental design has the properties of both experimental and non-experimental studies; part of the study may be non-experimental and the other part experimental.

**Sentence Completion Test:** Sentence completion test is a type of projective techniques of interviewing in which informant is asked by completing a given sentence to find association of the respondent with the particular topic of research.

**Sequential sampling:** *See complex sample design.*

**Short communications:** These are preliminary results of a project, perhaps one season's results, or results that are not of major significance but are nevertheless interesting. The exact nature of these communications will vary with the target publication.

**Short term trend:** Short term trend is the time series showing the direction of the series in a short period of time.

**Sign test for paired data:** *See two sample sign test.*

**Sign Tests:** The sign test is one of the easiest parametric tests which are based on the direction of the plus or minus signs on observations in a sample and not on their numerical magnitudes.

**Signed Rank Test:** *See Wilcoxon Matched-pairs Test.*

**Significance level:** The significance level indicates the likelihood that the answer will fall outside that range.

**Simple correlation:** *See Karl Pearson's coefficient.*

**Simple factorial design:** In case of simple factorial design, the effects of varying two factors on the dependent variable are considered.

**Simple hypothesis**: If a hypothesis is of the type $\mu=\mu_{H0}$, then the hypothesis is called as simple (or specific) hypothesis.

**Simple random sampling:** *See chance sampling.*

**Simple regression:** Simple regression is the regression between only two variables, one is defined as independent which is the cause of the behaviour of another defined as dependent variable.

**Simple tabulation:** The simple tabulation gives information about one or more groups of independent questions,

**Simulation approach to research:** It involves the construction of an artificial environment within which relevant information and data can be generated.

**Single factor ANOVA**: *See one-way ANOVA.*

**Skewness:** The skewness is a measure of symmetry or more precisely the lack of symmetry.

**Snowball sampling:** Snowball sampling is a process of selecting a sample using networks. To start with, a few individuals in a group or organization are selected using purposive, random or network sampling to collect the required information from them. They are then asked to identify other people in the group or organization who could be contacted to obtain the same information. The people selected by them become a part of the sample. The process continues till the saturation point in terms of information being collected.

**Sociometry:** The sociometry is a technique for describing the social relationships among individuals in a group.

**Software:** Software consists of computer programs written by the user which allow the computer to execute instructions.

**Spearman's rank correlation coefficient:** Spearman's rank correlation coefficient is a measure of association which is based on the ranks of the observations and not on the numerical values of the data. Because it was developed by Charles Spearman so it is known as Spearman's rank correlation coefficient.

**Spearman's rank correlation test:** *See Spearman's rank correlation coefficient.*

**Specific hypothesis:** *See simple hypothesis.*

**Stability aspect:** The stability aspect is concerned with securing consistent results with repeated measurements of the same person and with the same instrument.

**Stacked bar chart:** A stacked bar chart is similar to a bar chart except that in the former each bar shows information about two or more variables stacked onto each other vertically. The sections of a bar show the proportion of the variation they represent in relation to one another. The stacked bars can be drawn only for categorical data.

**Stakeholders in research:** Those people or groups who are likely to be affected by a research activity or its findings. In research there are three stakeholders: the research participants, the research worker and the funding body.

**Standard deviation:** Standard deviation is defined as the square-root of the average of the squares of deviations, when such deviations for the values of individual items in a data series are obtained from the arithmetic average. Standard deviation is most widely used measure of dispersion of a data series.

**Standard error:** The standard error (S.E) is the standard deviation of the sampling distribution of a statistic. It helps in testing whether the difference between observed and expected frequencies could arise due to chance.

**Statistic:** A statistic is working out of certain statistical measures from the samples to describe its characteristics.

**Statistical design:** The statistical design is concerned with the question of how many items are to be observed and how the information and data assembled and analyzed.

**Statistical test:** A statistical test is a formal technique, based on some probability distribution, for arriving at a decision about the reasonableness of an assertion or hypothesis.

**Statistics of attributes:** Descriptive data obtained on the basis of certain attributes are known as statistics of attributes.

**Statistics of variables:** Descriptive data obtained on the basis of certain attributes are known as statistics of variables.

**Stem-and-leaf display:** The stem-and-leaf display is an effective, quick and simple way of displaying a frequency distribution. The stem and leaf for a frequency distribution running into two digits is plotted by displaying digits 0 to 9 on the left of the $y$-axis, representing the tens of a frequency. The figures representing the units of a frequency (i.e. the right-hand figure of a two-digit frequency) are displayed on the right of the $y$-axis.

**Store audits:** *See distributor audits.*

**Story Completion Test:** The story completion test is a type of projective techniques of interviewing wherein the research worker may contrive stories and ask the respondents to complete them.

**Stratified random sampling:** Stratified random sampling is one of the probability sampling designs in which the total study population is first classified into different subgroups based upon a characteristic that makes each subgroup more homogeneous in terms of the class variable. The sample is then selected from each subgroup either by selecting an equal number of elements from each subgroup or selecting elements from each subgroup equal to its proportion in the total population.

**Stratified sampling:** Stratified sampling is a method that is applied if the population from which a sample is to be drawn does not constitute a homogeneous group. The items selected from each stratum are based on simple random sampling the entire procedure, first stratifying followed by simple random sampling, is known as stratified random sampling.

**Structured interviews:** Structured interview is the method of collecting information through personal interview carried out in a structured way. The interviewer follows a rigid procedure laid down, asking questions in a form and order prescribed.

**Structured observation:** Structured observation can be described by a definition of the units to be observed, the style of recording the observed information, standardized conditions of observation and the selection of data of observation.

**Structured questionnaire:** Structured questionnaire is one in which all questions and answers are specified and comments in the respondent's own words are held to the minimum. This is definite, concrete and contains pre-determined questions. The questions are presented with exactly the same wording and in the same order to all respondents. The form of the question may be either closed (of the type 'yes' or 'no' or multi choice) or open (i.e., inviting free response) but should be stated.

**Stub:** Stub is a part of the table structure. It is the subcategories of a variable, listed along the *y*-axis (the left hand column of the table). The stub, usually the first column on the left, lists the items about which information is provided in the horizontal rows to the right. It is the vertical listing of categories or individuals about which information is given in the columns of the table.

**Study design:** The term study design is used to describe the type of design that is going to be adapted to undertake the research study; that is, if it is going to be experimental, cor-relational, descriptive or before and after. Each study design has a specific format and attributes.

**Study population:** Every study has two aspects: study population and study area (subject area). People who are required to find out about are collectively known as the study population or simply populated and are usually denoted by the letter 'N'. It could be a group of people living in an area, employees of an organization, a community, a group of people with special issues, etc. The people from whom the information gathered, is known as the sample 'n', are selected from the study population.

**Subject area:** Any academic or practice field in which the research study is conducted, is called the subject or study area. It could be health or other needs of a community, attitudes of people towards an issue, occupational mobility in a community, coping strategies, depression, domestic violence, etc.

**Subject orientated scale:** This is a kind of scale that is designed to measure characteristics of the respondent who completes it or to judge the stimulus object which is presented to the respondent. This scale is used when a respondent scores some object without direct reference to other objects.

**Subjectivity:** Subjectivity is an integral part of the way of thinking that is 'conditioned' by the educational background, discipline, philosophy, experience and skills. Bias is a deliberate attempt to change or highlight something which in reality is not there, but it is done because of the vested interest. Subjectivity is not deliberate, it is the way to understand or interpret something.

**Summated rating scale:** *See Likert scale.*

**Summated scale:** This type of scale is investigated by using the item analysis approach in which a specific item is assessed on the basis of how well it separates between those respondents whose total score is high and those whose score is low.

**Surveys:** Surveys are concerned with describing, recording, analyzing and interpreting conditions that either exist or existed. The research worker does not manipulate the variable or arrange for events to happen. This technique of data collection is only concerned with the conditions or relationships that exist, opinions that are held, processes, going on effects that are evident or trends that are developing.

**Symbols:** Symbols are similar to abbreviations or acronyms, but they are usually shorter. symmetrical inkblots.

**System software:** *See operating software.*

**Systematic sampling:** Systematic sampling is a way of selecting a sample where the sampling frame, depending upon the sample size, is first divided into a number of segments called intervals. Then, from the first interval, using the SRS technique, one element is selected. The selection of subsequent elements from other intervals is dependent upon the order of the element selected in the first interval. If in the first interval it is the fifth element, the fifth element of each subsequent interval will be chosen.

**Systematic sampling:** Systematic sampling is to select the sample just after every pre-decided number. An element of randomness is usually introduced into this type of sampling by using random numbers to select the unit with which to start.

# T

**Table of random numbers:** Most books on research methodology and statistics have tables that contain randomly generated numbers. There is a specific way of selecting a random sample using these tables.

**Table:** A table is a systematic arrangement of data or information in a format that allows the reader to observe variations or trends and make comparisons. Tables offer a useful way of presenting analyzed data in a small space that brings clarity to the text and serves as a quick point of reference. There are different types of tables housing data pertaining to one, two or more variables.

**Tabulation:** Tabulation is a part of the technical procedure wherein the classified data are put in the form of tables.

**Tabulation:** Tabulation refers to the arrangement of mass of assembled data in some kind of concise and logical order. Consequently, the tabulation is the process of summarizing raw data and displaying the same compact form for further analysis.

**T.A.T.:** The T.A.T. or Thematic apperception test is a type of projective techniques of interviewing containing a set of pictures that are shown to the respondents and are asked to describe what they think the pictures represent.

**Technical report:** A technical report may be a preliminary report of a piece of research that is interesting, but suitable or intended for peer-reviewed publication. In many cases, a working paper might be created later, with the expansion of more material, in a scientific paper.

**Technical writing:** Technical writing deals with the topics of more technical nature. It concerns with the areas of science and technology.

**Telephone interview:** Telephone interview is the method of collecting information which consists in contacting respondents on the telephone itself. The method is not a very common, but plays an important part in industry surveys, particularly in developed regions.

**Terminology:** Technical terms and words or the phrases having a particular and special meaning may be defined as terminology.

**Test statistic:** The test statistic is the value worked out from the sample data that correspond to the parameter under investigation.

**The quantitative approach to research:** The quantitative approach to research includes the generation of data in quantitative form which can be subjected to rigorous quantitative analysis in a formal and rigid fashion.

**Thematic Apperception Test:** *See T.A.T.*

**Thematic writing:** A style of writing, which is written around main themes.

**Theoretical framework:** As reading the literature is started, it will soon discover that the problem that is wished to investigate has its roots in a number of theories that have been developed from different perspectives. The information obtained from different sources needs to be sorted under the main themes and theories, highlighting agreements and disagreements among the authors. This process of structuring a 'network' of these theories that directly or indirectly has a bearing on the research topic is called the theoretical framework.

**Theory of causality:** The theory of causality advocates that in studying cause and effect there are three sets of variables that are responsible for the change. These are-cause or independent variable, extraneous variables and change variables. It is the combination of all three that produces a change in a phenomenon.

**Thesis:** *See Dissertation.*

**Thurston scale:** The Thurston scale is one of the scales designed to measure attitudes. Attitude through this scale is measured by means of a set of statements, the 'attitudinal value' of which has been determined by a group of judges. A respondent's agreement with the statement assigns a score equivalent to the 'attitudinal value' of the statement. The total score of all statements is the attitudinal score for a respondent.

**Thurston-type Scale**: *See differential Scale.*

**Time series:** Series of successive observations of the given phenomenon over a period of time are referred to as time series.

**Tomkins-Horn picture arrangement test:** The Tomkins-Horn picture arrangement test is a type of projective techniques of interviewing that is designed for group administration. It consists of several plates, each plate contains sketches that may be arranged in different ways to portray the sequence of events.

**Total sum of squares:** When Eigen values of all factors are summed, the resulting value is called as the total sum of squares.

**Transferability:** The concept of transferability refers to the degree to which the results of qualitative research can be generalized or transferred to other contexts or settings.

**Transitory consumer panel:** A transitory consumer panel is set up on consumer panel to measure the effect of a particular phenomenon.

**Treatments:** The different conditions under which experimental and control groups are put are usually referred to as treatments.

**Trend curve:** A set of data measured on an interval or a ratio scale can be displayed using a line diagram or trend curve. A trend line can be drawn from data pertaining to both a specific time and a period. If it relates to a period, the midpoint of each interval at a height commensurate with each frequency is marked as a dot. These dots are then connected with straight lines to examine trends in a phenomenon. If the data pertains to an exact time, a point is plotted at a height commensurate with the frequency and a line is then drawn to examine the trend.

**Trend studies:** These studies involved selecting a number of data observation points in the past, together with a picture of the present or immediate past with respect to the phenomenon under study, and then making certain assumptions as to the likely future trends. In a way compiling a cross-sectional picture of the trends is being observed at different points in time over the past, present and future. From these cross-sectional observations conclusions are drawn about the pattern of change.

**t-test:** The test is based on the assumption of normality. The t-test is based on t-distribution and is considered for judging the significance of a sample mean or for judging the significance of difference between the means of two samples in case of small sample(s) when population variance is not known i.e. in which case variance of the sample is used as an estimate of the population variance.

**Two sample sign test:** The sign test has an important implementation in problems, where paired data are dealt. In such problems, every pair of values can be replaced with a plus sign (+) if the first value of the first sample (say X) is greater than the first value of the second sample (say Y) and minus sign (–) if the first value of X is less than the first value of Y. In case the two values are similar, the concerning pair is discarded. In the case, two samples are not of similar size, then some of the values of the larger sample left over after the random pairing will have to be discarded. The testing technique remains as such started in the case of one sample sign test.

**Two-factor-factorial design:** *See simple factorial design.*

**Two-state devices:** The transistors on an IC Chip take only two states, e.g. either they are on or off, conducting or non-conducting.

**Two-tailed and One-tailed tests:** A two-tailed test rejects the null hypothesis, when, the sample mean is significantly higher or lower than the hypothesized value of the mean of the population. Symbolically, the two tailed is presented as $H_0$: $\mu=\mu_{H0}$ and $H_\alpha$: $\mu\neq\mu_{H0}$, which means either $\mu>\mu_{H0}$ or $\mu<\mu_{H0}$.

**Two-way ANOVA technique:** Two-way ANOVA technique is employed when the data are classified on the basis of two factors.

**Type I error:** In testing a hypothesis, many reasons may sometimes be commiting a mistake and drawing the wrong conclusion with respect to the validity of the hypothesis. If a null hypothesis is rejected when it is true and should not have been rejected, this is called a Type I error.

**Type I errors:** Type I error means rejection of hypothesis which should have been accepted.

**Type II Error:** In testing a hypothesis, for many reasons may sometimes be committing a mistake and drawing the wrong conclusion in terms of the validity of the hypothesis. If a null hypothesis is accepted when it is false and should not have been accepted, this is called a Type II error. In other words Type II error means accepting a hypothesis which should have been rejected.

# U

**Uncontrolled observation**: Uncontrolled observation is the observation when it takes place in the natural setting and the situations cannot be controlled by the research worker.

**Unethical:** Any professional activity that is not in accordance with the accepted code of conduct for that profession is considered unethical.

**Uni-dimensional scale:** The uni-dimensional scale measures only one attribute of the respondent or object.

**Uni-variate population:** Uni-variate population is the population consisting of measurement of only one variable.

**Universe:** The universe means to the total of the items or units in any field of inquiry.

**Unstructured interview:** Unstructured interview is characterized by a flexibility of approach in questioning. The procedure does not follow a system of pre-determined questions and standardized techniques of recording information. In this the interviewer is allowed much greater freedom to ask supplementary questions if needed or omit certain questions if the situation so requires. Interviewer may also change the sequence of questions.

**Unstructured observation:** Unstructured observation is the observation that is not described by a careful definition of the units to be observed, the style of recording the observed information, standardized conditions of observation and the selection of pertinent data of observation.

**Unstructured questionnaire:** *See Non-structured questionnaire.*

**U-test:** This is a very popular, rank sum test used to determine whether two independent samples have been drawn from the same population. It uses more information in comparison to sign or Fisher-Irwin test. This test applies under very general conditions and necessitates only that the populations sampled should be continuous.

# V

**Validity:** The concept of validity can be applied to every aspect of the research process. In its simplest form, validity refers to the appropriateness of each step in finding out what is set out to. However, the concept of validity is more associated with measurement procedures. In terms of the measurement procedure, validity is the ability of an instrument to measure what it is designed to measure. Validity is the most critical criterion to indicate the degree to which an instrument measures what it is supposed to measure. In other words, validity is the extent to which differences found with a measuring instrument reflecting true differences among those being tested.

**Variable:** A concept which can take on different quantitative values is called a variable. An image, perception or concept which can be measured and able to take different values, is called variable or in other words, a concept that can be measured is called a variable. A variable is a property that takes on different values. It is a rational unit of measurement that can be assumed any one of a number of designated sets of values.

**Variance:** Variance is an important statistical measure and is described as the mean of the squares of deviations taken from the mean of the given series of data. It is a frequently used measure of variation. In simpler way it can be defined as the square of standard deviation.

**Verbal Projection Test:** The verbal projection test is a type of projective techniques of interviewing wherein the respondent is asked to comment on or to explain what other people do.

**Warranty cards:** Warranty cards are generally posted sized cards which are used by dealers of consumer durables to collect information regarding their products.

**Wilcoxon Matched-pairs Test:** The Wilcoxon Matched-pairs Test can be used if the research situations where two-related samples, i.e. case of matched pairs such as a study where husband and wife, or the output of two similar machines or some subjects are studied in context of before-after experiment, the magnitude of difference between matched values in both the directions are to be determined. While applying this test, first the differences (di) between each pair of values need to be found followed by and assigning rank to the differences from the smallest to the largest without regard to sign. The actual signs of all difference are then put to corresponding ranks and the test statistic 'T' is calculated.

**Wilcoxon-Mann-Whitney test:** See U-test.

**Word Association Test:** Word association tests are a type of projective techniques of interviewing that is used to extract information regarding such words which have maximum association.

**Working definition:** *See Operational definition.*

**Working hypotheses:** Working hypotheses are a set of suggested tentative solutions of explanations of a research problem which may or may not be the real solutions.

**Working papers:** *See Technical report.*

# X

**$\chi^2$-test:** The $\chi^2$ test (chi square test) is based on the assumption of normality. The $\chi^2$-test is based on chi-square distribution and as a non-parametric test, it is used for comparing a sample variance to a theoretical population variance.

# Y

**Yates' correction:** Yates has suggested a correction for continuity in $\chi^2$ value calculated in a 2×2 table, predominantly when cell frequencies are small and $\chi^2$ is just on the significance level. The correction suggested is known as Yates' correction, which involves the reduction of the deviation of observed from expected frequencies that reduces the value of $\chi^2$.

# Z

**z-test:** The z-test is based on the assumption of normality. The z-test is based on the normal probability distribution and is used for judging the significance of several statistical measures, particularly the men.